To Freddie with all my love
Christmas 1997 Bab

CRUISING YACHTS:
Design and Performance

T. Harrison Butler

CRUISING YACHTS:
Design and Performance

T. HARRISON BUTLER, AINA

Foreword by Ian Howlett, C. ENG MRINA

Seamanship, like any other form of skill, is an art and cannot be
pursued at odd times as a secondary occupation;
on the contrary, no other work may be subordinated to it.
THUCYDIDES, Book I: CXLIII.

EXCELLENT PRESS
London

First Published 1945
Revised Second Edition 1947
Third Edition 1958

Fourth Edition published 1995 by Excellent Press,
103 Lonsdale Road, Barnes, London SW13

ISBN 0 8548 7397 0

Printed in Great Britain by
BAS Printers Ltd, Over Wallop, Hampshire

CONTENTS

Acknowledgments to the Fourth Edition

THIS edition results from the requests, insistence and cajolery of family and friends (all members of the Harrison Butler Association) and, finally, the time limit set by David Burnett.

My thanks to the editors of *Yachting Monthly*, *Yachting World*, *Yachtsman* and *Motor Boat* for permission to use material from those publications in this new edition.

Photographs have been culled mostly from THB's collection and some of the sources are unknown. If we have inadvertently reproduced somebody's copyright material, I hope that it will be regarded as a compliment.

I have had particular help from Peter Hasler, Brian Hawkins, Ian Howlett, Peter Mather and my sons Timothy and Colin with research, criticism, word-processing, advice on the choice of designs and photographs and general encouragement; and from David Burnett and Sue Coley with the practicalities of publishing.

THB deserves the best we could contrive between us.

O.J. J-B

LIST OF PLATES

9 *Merrythought* (19′ LWL), Zyklon design. Built 1938 in Plymouth for C.F.D. Roberts.
Grey Owl under construction.

10 *Vindilis* (22′ 6″ LWL). Davinka/Vindilis design. Built 1935 by A.H. Moody & Son for THB and launched by EB.
Vindilis under power.
The table in use.
Interior, looking aft.

11 *Vindilis* under sail in the Beaulieu River; OJB and THB in the English Channel, 1937.
Interior, looking forward.

12 *Vindilis*: EB and THB during their last cruise, in 1939.
At anchor in the Medina River, Isle of Wight, as it used to be.

Foreword to the Fourth Edition

MY first real contact with the work of HB came not through an earlier edition of this book but on the ancient wooded riverwalk that makes its way from Beaulieu to Bucklers Hard. Along this beautiful and unspoilt river there lie, on swinging moorings, the inevitable selection of modern production craft, but just off the jetty by Keeping Trees was then stationed a lovely, azure blue, traditional cruising yacht with the name of *Ardglass*. Admiration for her proportions and form led to conversations in the laying up season with the owner, Pat Russell, as he gave her planks their winter treat of diesel. It transpired that this 6-tonner had been constructed for his father in 1929 to HB's 'Cyclone II' design, since when she had been an integral part of that family.

Acquisition of a first copy of the book led to great respect for a Corinthian designer with a special sense of form and a unique style. A few years later with the departure of *Ardglass* for the West Country, we were fortunate enough to take over that same mooring and in her place substitute a younger sister, *Sabrina*, a 1935 construction by Clemens to HB's 7-ton 'Yonne' design. She had previously looked after and been home to but two single-handing owners. These boats are for life it seems . . . indeed, the first owner of *Ardglass* was to expire attempting to coax life into her ancient 'Iron Topsail' at the mouth of the Beaulieu.

All HB designs, which were considered beamy by the standards of the time but not by those of today, were sculpted around eminently sensible midship sections. The shapes appear to be direct and highly refined descendants of the Falmouth Quay Punts which were noted for their handiness and weatherly qualities.

Generous headroom and space is combined with moderate freeboard and achieved via depth of canoe body and accompanying displacement. Displacement or rather lack of the same is very much a modern rule-bred fad, born of the racing success of offshore dinghies which rely on crew weight rather than proper ballast for adequate stability, and have no real requirement for accommodation. Average speeds in cruising seem however to be largely a function of performance in light airs and displacement need be no disadvantage in the same, so long as the craft is not undercanvassed. Transocean experiences such as those of *Askadil* and *Watermaiden* would seem to endorse this view.

Doubtless, the quest for perfect balance under sail must always be eternal, but hull volume balance seems essential to the good manners of yachts with keel-hung rudders. Turner's Metacentric Shelf system is little spoken of today and generally regarded as being only of academic interest; indeed, balance appears to have become again the Cinderella of design. Rather simpler, 'shift of LCB with heel' systems are however used to good effect by model yachtsmen and wiser full size practitioners. One eminent model yacht designer imposes a limit of movement of LCB from upright to 22.5 degrees of heel of 0.2% for every 1% the upright LCB lies behind the midsection and this method, which is derived purely from experience, certainly produces models with remarkable characteristics, as of course did the Metacentric system.

The balance of HB's later designs brought with it an increase of bow angle. The 'Omega' design has a half angle of entrance some 3 degrees wider than the 'Yonne' and as may well be considered appropriate, my own preference is for the earlier design! It may just be that the accompanying increase in rake of the keel, produced as powerful an effect as the change in volume distribution.

The traditional structure of 'husky cruising yachts' of this era was invariably heavy but that overbuild has generally stood them in good stead. The 'Yonne' design, with a displacement/length ratio of 482, is said to have 2.2 tons of ballast in 5.5 tons, a ratio of 40%. With modern but unexotic materials a ratio of 60% could be readily achieved, with a significant increase in stability. HB himself was very much in favour of 'scantlings cut down to racing practice, or nearly so'. 'Nearly so' is however a wise cautionary note in an age when 80ft racers have scantling weights little greater than those appropriate to dinghies.

Spars might of course also be considerably lighter in aluminium alloy or composites, but here again ultimate seaworthiness (as well as roll motion) is often enhanced by increased transverse inertia and model tests without masts in breaking seas have shown that a full 360 degree roll is far more likely in such circumstance.

Today's yachtsmen and yachtbuilders seem largely to have forgotten the romantic and aesthetic aspect of their sport. Perhaps in this age of mass production GRP such things are more remote and boats, like those detained at Her Majesty's pleasure, have in the dinghies at least, become numbers rather than names. The satisfaction of traditional wood construction, truly a living, albeit demanding structure, is itself remarkable; to lie in the foc's'l, look up and inspect grown knees, oak beams and laid decks is a therapy known only to those fortunate enough to find relaxation in such craft.

HB designs are certainly well travelled and dispersed throughout the world, but even in the most remote of ports they are instantly recognisable to those with an eye for a proper yacht. The term 'Classic' has been much overused of late, but seems a thoroughly correct term for all these designs.

The Association of HB yachts (owners are considered but caretakers) which has been so carefully nurtured over the years by HB's daughter, Joan Jardine-Brown, has enabled many owners to understand and appreciate their craft more fully, as well as enjoy the company of fellow trustees.

This Fourth Edition of *Cruising Yachts: Design and Performance* has been expanded to include many more designs and related information. The book, quite apart from its entertainment and instruction, provides an illuminating window into cruising, the way it was, when time ran slower and the Marina had yet to be.

The work is a gem and will give pleasure and inform countless numbers of today's yachtsmen who will only regret that they cannot have the pleasure of meeting HB in person. What a pity that time did not permit him to write the other two intended works.

Ian Howlett CEng MRINA
Oxford 1994

AUTHOR'S PREFACE

THE major and most important part of this book was published in *The Yachtsman* as a series of articles on *Simple Yacht Designing*. I received several requests that they should be reproduced in book form, so I have ventured to add one more volume to our yachting literature. It has also been suggested to me that I should produce a volume of some of my designs. My answer is that none of my premetacentric designs satisfy my present ideas of what a yacht should be, and that most of the more recent ones are capable of improvement. But I have embodied in this work some of my latest drawings and I hope that they represent yachts that have reached finality. One of the designs, *Prima*, is the work of my youngest daughter, Mrs. O. J. Jardine Brown, and I think that it is probably the best of the lot. This design is of some interest, for I know of only two others designed by a woman that have appeared in the yachting journals. They are both the work of a girl who in 1901 was Miss Maimie Doyle, the daughter of H. E. Doyle, the yacht-builder in Kingstown. The larger one, *Granuaile*, was built by Doyle and proved a great success. I am also indebted to my daughter for the pretty little sketch of a section being poised on a razor (page 36).

I have during the past few years received a large number of letters asking for information about metacentric analysis. For this reason and for others, I have paid considerable attention to this recent advance in naval architecture. Many additional chapters have been added to the original and so I have expanded the title to include performance as well as design. I hope that I have made the art and science of yacht design as simple as possible, and reduced the calculations to their simplest forms; where possible I have "poised" rather than calculated. The razor blade has replaced arithmetic, and all the necessary computations that remain can be accommodated on a post-card. Whereas much of the work has already been printed and then amplified, I fear that there will be some repetition, and that information that should have been introduced early in the book will be found in the chapter headed "Odds and Ends." I have found this "dust-bin" chapter most useful, for anything that I had forgotten and thought of from time to time was just pitchforked into the bin, if one can use a pitchfork for this purpose.

I have adopted a more intimate style than is usual in a strictly scientific book, but perhaps it is not really scientific, and I hope that in this way

I may have made the book more readable. It may be objected that I have given too much weight to metacentric analysis, but it seems to me that Admiral Turner's researches mark an entirely new epoch in yacht design, and that his methods are revolutionary in that they teach us how to fashion a hull that has an inherent tendency to run straight when heeled. Although we cannot say that a yacht with a good analysis will *ipso facto* be a good performer, we can say that one with a bad analysis will almost certainly be a bad yacht. I am also well aware that many of our younger designers, and at the moment we have a wealth almost of genius, are deeply interested in hull-balance and will welcome my attempts to throw some light on to a dark problem. I may even hope that some of them will be stimulated to further research, and I suggest that some of this should be carried out with models whose form must be untrammelled by racing formulae.

I have received great help and encouragement from many friends and from the readers and staffs of our prominent yachting journals. To these I render my thanks.

I am deeply indebted to Engineer Rear-Admiral Alfred Turner, whose work and personal instruction have given me an entirely new outlook upon yacht designing. To him I owe not only my initiation into the mysteries of hull-balance, but also the instruction he gave me in simple methods of calculation and the method of "poising." I thank him for his friendly criticism of many aspects of my work, and for the valuable constructive suggestions he has made in my explanations of his own researches.

Lieut.-Commander Braithwaite, R.N.V.R., has contributed a most valuable chapter explaining the methods of finding the centre of gravity of a yacht, a subject that has been glossed over or even neglected in many books on yacht designing and construction.

Mr. K. Adlard Coles and Mr. Eric C. Hiscock have given me great encouragement and valuable help; in fact Mr. Coles went so far as to extract the articles from *The Yachtsman* to sew them together, and hold them before my nose like a carrot before a donkey, saying: "Now get on with it."

I thank the Editor of *Yachting Monthly* for allowing me to make free use of my original article on metacentric analysis that appeared in his journal and for permission to use other matter. The Editor of *Yachting World* has also allowed me to make use of articles that have appeared in his journal and has most kindly lent me some of the blocks that illustrated these articles.

I must not forget the debt that I owe to Mr. S. J. Aspinal of the firm of J. A. Reynolds & Co., Birmingham, for the splendid tracings that he has made of many of my drawings. Most of these were drawn on hot-pressed Whatman paper, and for the purposes of blockmaking a true-to-scale reproduction is necessary on tracing cloth. I have not time for tracing and Mr. Aspinal has done the work far better than I could have done it. Much of the success of the reproductions is due to him and I thank him for his care and great skill.

Finally, I hope that my readers will excuse the many faults and omissions in my book and will realize that I have had to work in the little spare time that my war work has left me. I am not a professional designer, but an ophthalmic surgeon, an occupation that in these days is very exacting.

T. HARRISON BUTLER, A.I.N.A.

CRUISING YACHTS: DESIGN AND PERFORMANCE

CHAPTER I

General Considerations

THERE are many who have a general idea of a yacht in their minds, a mental picture, that they would like to develop on paper, and see the finished design before their eyes. There is absolutely no reason why this wish should not be gratified, and it will be my object to explain how this ambition may be achieved in the simplest possible manner. All that I know about the subject I have learned from books and by careful study of actual yachts and models. I began at Oxford by going to the Radcliffe Camera and studying Dixon Kemp's well-known work on Yacht Architecture, but I owe a debt of gratitude to the Glasgow naval architect, Mr. J. Paterson, who contributed a series of articles to *The Yachtsman*, which will be found in Vol. X, January, 1896, page 157. These were anonymous, but internal evidence clearly showed who was responsible for this masterly work. A still more valuable series by the late Albert Strange commences in *The Yachting Monthly*, Vol. XVII, page 257. My only excuse for writing this book is that the far better productions mentioned are not now available to all.

Many are deterred from beginning to design by thinking that most elaborate mathematical calculations are involved, or that one must be a competent draughtsman. Simple methods have so reduced the calculations that they can be accommodated on a post-card. In any case they involve only addition, subtraction, multiplication and division. It is not at all necessary to be a finished draughtsman, although if the design is to be published, a presentable drawing is necessary. What then is required? First common horse sense; then artistic feeling with a flair for fair curves is helpful; and finally untiring patience and a determination to attain perfect accuracy is indispensable. The old proverb "A lazy man takes most trouble" is doubly true in yacht designing.

I am often asked "How do you begin a design?" One begins it in one's bath; at night, "when sleep her balm denies"; and I regret to say perhaps during sermons! I have the whole design clearly in my head before I put anything on paper. I can see the curves in my mind's eye and gradually mould them into harmony and symmetry. Then if the design is to be entirely new I should take any good bit of paper, and with a suitable scale, put my ideas down, chiefly freehand. All that is necessary is a few simple curves and a pair of dividers. Fig. 1 shows a drawing of this kind which was 7 in. long. I first decided that the water-line was to be 18 ft., and using a scale of $\frac{1}{4}$ in. to 1 ft. ruled a water-line and marked on it two points 18 ft. apart, actually $4\frac{1}{2}$ in. The draft was to be 4 ft., and the least freeboard one-tenth of the L.W.L. I now sketched in the yacht as I thought she ought to be, as seen from the side. This would be a *side elevation* to the architect but we call it the *sheer plan*. The beam was to be 7 ft., and so I drew a half mid-section $3\frac{1}{2}$ ft. wide. Above this I drew another line parallel with the water-line, and sketched in the deck-plan. The advantage of this small drawing is that it is an effort of the eye, and there is no slavish dependence upon a mechanical curve. Then the small sketch can be taken in at a glance, and an opinion formed regarding the general appearance of the design. This drawing has so far got no further, but perhaps one day she may be finished and become the *Sprite-of-Arden*.

Commencing with a sketch of this kind, the next step will be to turn it into a proper design, and for this purpose a little apparatus is needed. First a good drawing board accurately made. One that is inaccurate is a constant source of trouble and may lead to a faulty drawing. I like to have two boards; a large one about 28 in. by 32 in. which will comfortably take Elephant drawing paper, and a small one, about 22 in. by 16 in. This is useful for the sail-plan, and for making a preliminary half-scale drawing of the design. It is far easier to rough out the design on a small scale and then double it for the final fairing up and completion. I am visualizing the small quarter-scale preliminary free-hand sketch, then the half-scale accurate drawing, and finally the full-scale design. This may seem a time-wasting method, but for a beginner it is not so. There will be hundreds of alterations to be made, and perhaps one or two complete drawings, and all this is easier on a small scale. I personally never use the large scales that the professional designer would employ. Up to say a 20-ft. water-line I work to a scale of 1 in. to 1 ft., and then I use a scale of $\frac{3}{4}$ in. to 1 ft. After 30 ft. on the L.W.L. I work to a scale of $\frac{1}{2}$ in. to 1 ft. This brings the drawing on to the Imperial size of paper. And now we will consider the paper.

It is waste of money and labour to use poor paper. Buy the most expensive kind that is made, hot-pressed Whatman, and if possible choose a sample that is not new but has matured. At one time I could not imagine why I could not fair up my drawing correctly, and I was astounded to find that my sections and water-lines were not equally spaced, although I knew that I had been most careful to get them correct. Then I discovered that a water-line which on one day was 20 in. long was on the next day $\frac{1}{8}$ in. longer! Yes, paper swells and swells unequally with changes in the humidity of the atmosphere. Matured Whatman does not alter as much as newly made. but I fear that it is not possible to get paper as good as it used to be. Keep the drawing board away from an outer wall, and keep it as dry as possible. To avoid inaccuracy try to rule in all the sections and water-lines at one sitting, and in any case check them up occasionally. *If the spacing of the water-lines, sections and buttock. lines is not absolutely accurate it will be impossible to fair up the design properly-* In addition to drawing paper, it will be necessary to have some tracing paper and tracing cloth. I buy the very best detail paper for tracing. A few instruments are needed and some curves and a set of scales.

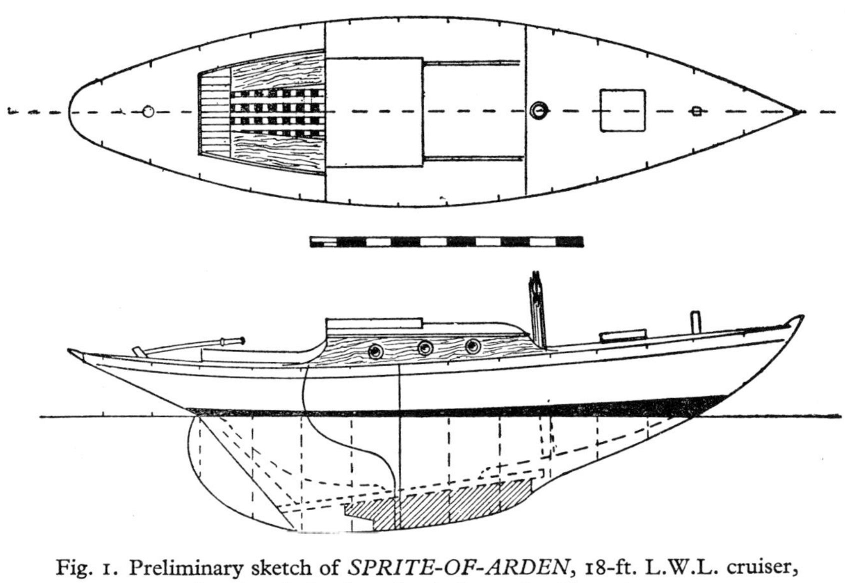

Fig. 1. Preliminary sketch of *SPRITE-OF-ARDEN*, 18-ft. L.W.L. cruiser, showing deck-plan, sheer-plan and mid-section.

The instruments are a pair of dividers, compasses for ink and pencil, and an inking-pen. I have a small pair of inking-compasses, but my pencil pair cost threepence and has served me well for thirty years. It just happens to have a perfect hinge or joint. It is one of the kind that children use and I put in my ends of pencils. Albert Strange hooted with mirth when he saw them and called them my "tongs."

Buy the best pencils. I use an HB for tracing paper; an H for ordinary work, and an HHH for drawing on the Whatman. I am well aware that I ought to sharpen them to a chisel edge, but I fear that I use the ordinary point. It is nice to have a mechanical sharpener, for there is constant pointing to be done. The inking-pen must be the best that can be bought (the German Reeflex is the best). It should open out without altering the spacing between the points so that it can be cleaned *every time it is inked*. One does *not* dip the inking-pen into the ink; the ink is inserted with the quill that is found attached to the stopper. I think that Higgins' is the best. I have tried Pelikan ink in tubes, but I found that after a time it got too thick and did not run well. But this brand is excellent for colours.

A large and a small T-square are necessary, and the large one should have an ebonite edge. Two or three set-squares are required, one a 45 degree and another a 60 degree. I think a large and a small size in each is desirable. Then we need some instruments peculiar to yacht designing, some curves, splines and weights. Whatever you have or have not, a set of Dixon Kemp's "pears" is indispensable. They can be got from Stanley's in London, or from J. A. Reynolds & Co., Edmund Street, Birmingham. Two or three splines are necessary. One a thickish type for sheer lines and two thinner for water-lines, and such like. I believe that celluloid are said to be the best, but I use the ordinary lance-wood pattern. I object to celluloid squares, curves and splines because they become electrified and pick up small particles of india-rubber which are most tenacious. These are deadly in that they guide the ink under the spline and a nasty mess is the result.

Five lead weights are necessary. These cost rather a lot of money but are quite easy to make. They are used to hold the splines in place. And now for a shock! Unless the designer is going to make one design every three years he will need a planimeter. It may, in fact it must be, possible to obtain one second-hand, but if new the simple type measuring to a scale of an inch to a foot only, costs or did cost four to five pounds. Personally, if I had to go without this instrument I would give up designing, for the necessary measurements are most wearisome. A slide-rule is useful, but not necessary.

To return for a moment to curves. Years ago I bought a box of naval architect's curves and I thought that I was going to have a treat, but alas, there was not an accurate one among them. It is almost impossible to buy accurate specimens. They have false curves and flats. Go to an instrument shop and select a few with the very greatest care. I have one about 10 in. long that is worth its weight in gold. I use it for buttocks and other longish curves, and if I were to lose or break it I should have lost one of my dearest friends. It is for this reason that I recommend the half-scale preliminary drawing. It is possible to do a lot of freehand drawing and to see at once when a curve is false or has flats on it. Personally I abominate flats even in the body plan, but one often sees them in the design of a bow. Let us have free flowing curves with the reverses melting into the main curve like the curve of a flower. The circle and the parabola are to be aimed at, in fact some have designed yachts by a system of co-ordinate geometry and then naturally every curve must be an harmonious one.

Having our preliminary sketch and our instruments we are ready to begin. I never even think of any calculations till the design is complete; I design entirely by eye. When the design lies before me faired up, then I put it to the test; weigh it, adjust the lead keel and work out its metacentric balance. Of course alterations may be necessary and a new drawing be called for, but if you are not prepared for trouble leave designing alone. Most of my early designs were drawn and re-drawn in some cases three or four times till I was satisfied. I know one designer, and one of great experience, who made nine completed designs, all most beautifully finished, before he obtained his final and perfect result.

Having made the preliminary sketch and decided upon the chief dimensions and the general form, we then proceed to fill in more details, but before we do so it will be wise to consider some important principles. The keystone of a design is the midship section. I have dealt fully with this subject in *The Yachting Monthly*, Vol. LXIII, page 35, and I do not intend to say more about the subject here than is absolutely necessary. Many a fine drawing has been ruined by an unfortunate choice of mid-section. Two points of view have to be considered and to some extent they are antagonistic, and a compromise is necessary. A cruising yacht is in the first place a sailing machine and in the second a home. In the case of a day boat we can have any kind of mid-section that we think best, but if we have a cruising yacht in mind we must choose a section that will give us sitting-room in the smaller sizes and standing-room in larger vessels, of course in this case taking the cabin top into consideration. It will, therefore, be wise at this stage to draw a mid-section that includes the construction and is a section of the cabin-plan. Fig. 2 is an example of such a section. It is the mid-section of the yacht *Sinah* that will be utilized to illustrate this book. She is 24 ft. on the load water-line and has a beam of 8 ft. 6 in. Within this mid-section we obtain good sitting-room over the cushions, and under the deck-beams, and 6 ft. head-room with a cabin top of moderate height. In a small yacht, say a 4-tonner, we must of course have the same sitting-room, but it cannot be under the deck-beams. Either the cabin top must be brought further out, or the top-sides must be raised to give the necessary height. Standing-room is, of course, impossible. We see now that our mid-section is to some extent predetermined; we cannot make it any shape we like, except in a large yacht or a day boat. But it is the worst possible practice to build a yacht round a cabin-plan. An owner may ask for this, but the naval architect must be very firm with him, and insist upon a sailing yacht not a moving houseboat. The shape of the mid-section determines the stability of the hull. To explain this I must touch upon those principles of hydrostatics that govern the stability of a hull. Any body floating in water displaces its own weight of water. The weight of this water pushes the hull upwards and its own weight pushes it down. The two forces neutralize each other and equilibrium is established. We can assume that the upward acting force acts through the centre of gravity of the displaced water, a centre that we call the centre of buoyancy. We shall learn later how to determine the position of this point. In the same way the downward acting force acts through the centre of gravity of the hull. The centre of gravity is a fixed point which cannot alter unless ballast is added or removed, but the centre of buoyancy moves laterally when the hull is heeled. Fig. 3 shows the mid-section of *Rory*, an ocean cruiser designed by Rear-Admiral Alfred Turner. The centre of gravity, in this case situated below the centre of buoyancy is seen at G, and the centre of buoyancy at B. When the yacht heels the centre of buoyancy moves out, and in the drawing it is shown at the position it occupies at an angle of

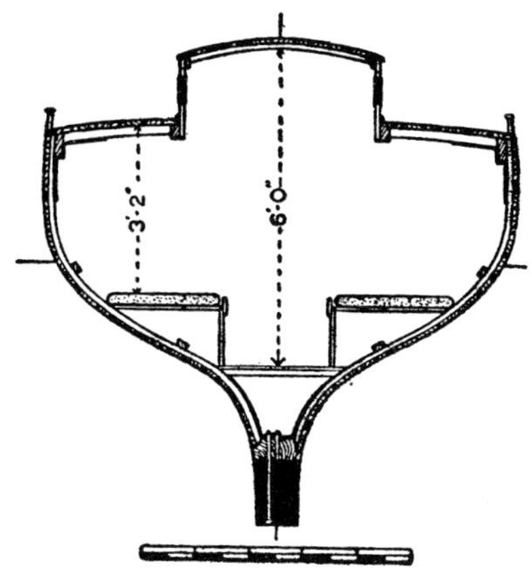

Fig. 2. Mid-section of *SINAH*.

30 degrees, labelled B^1. From B^1 we have erected a perpendicular to the inclined water-line which cuts the centre line at M^1. This is a point called the metacentre, or rather the inclined or prometacentre.

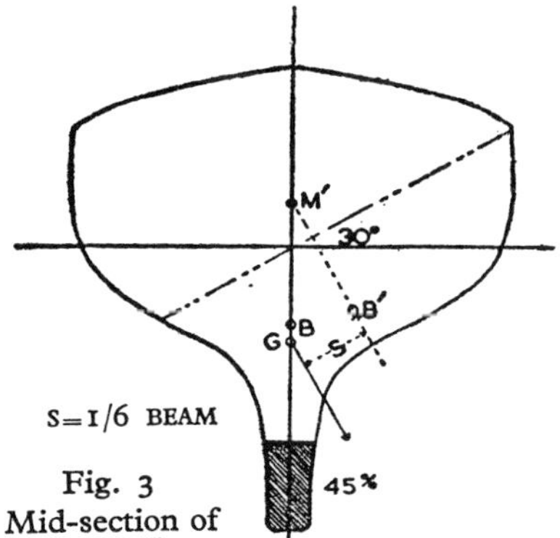

Fig. 3
Mid-section of
RORY

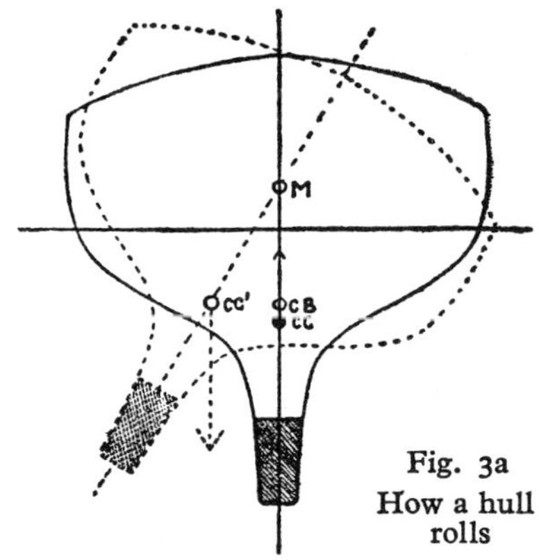

Fig. 3a
How a hull
rolls

Fig. 3a illustrates Admiral Turner's conception of how a hull rolls. The original vertical axis containing the metacentre M and the centre of buoyancy CB does not move. If, as in this case, the hull is a metacentroid, she swings on her metacentre as shown and the centre of gravity swings out to CG^1. The centre of buoyancy does not swing out but pulsates a little up and down. The metacentre also pulsates in the same manner.

In the case of this particular yacht the inclined metacentre coincides with the actual metacentre, and hull is a metacentroid.* A study of the drawing will show that the upward pressure of the displaced water acting in the direction of the arrow, is trying to bring the hull to an even keel, while the total weight of the hull acting downwards in the direction of the second arrow, is having the same effect. There is a righting couple acting by means of the assumed lever S. The stability of the yacht then depends upon its total weight and the length of the lever S.

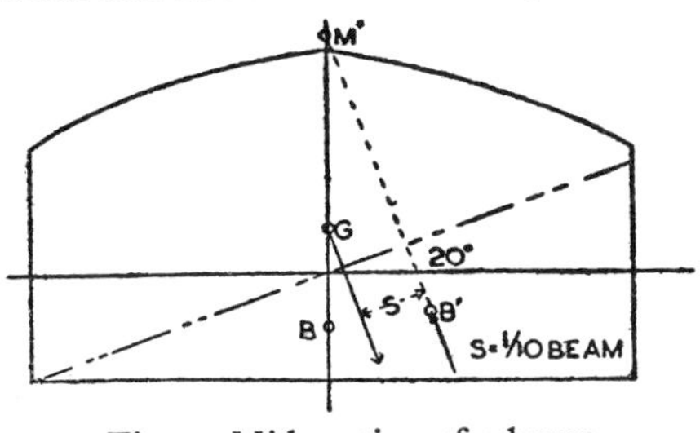

Fig. 4. Mid-section of a barge.

In the case of *Rory*, S is equal to one-sixth of the beam, a high figure. Its length is largely due to the fact that *Rory* carries 45 per cent. of her total weight in her lead keel. A mid-section that allows the centre of

* For Turner's definition of a metacentroid see page 108 *post.*

buoyancy to move out rapidly, associated with a low centre of gravity, gives the highest stability. The former effect is attained by having a flat floor and hard bilges.

Fig. 4 is the mid-section of a barge. Although heeled to only 20 degrees the centre of buoyancy has swung a long way out, but S is equal only to one-tenth of the beam. This is because the centre of gravity G is high up, and above the centre of buoyancy. It now we combine the section of a barge with a fin keel bringing the centre of gravity low down, we shall have the most stable mid-ship section that is possible (Fig. 5). We have rounded off the edges of the bilges, and arrived at the mid-section that was so popular and so efficient in the days of the length and sail area rating rule.

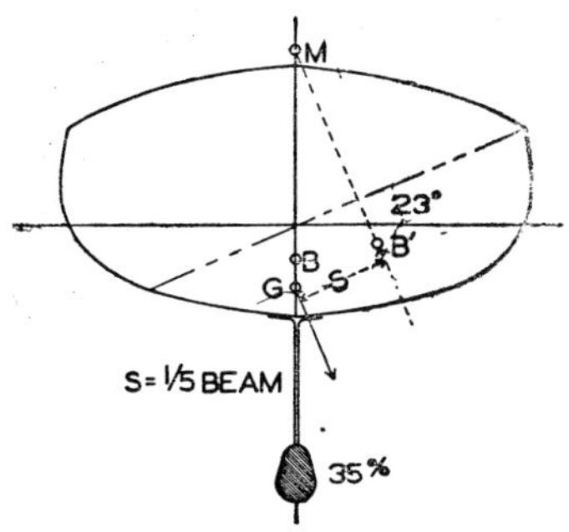

Fig. 5.
A stable mid-section.

The actual formula was length on the L.W.L. multiplied by the sail area divided by 6,000. It was simple and practical in the small classes, but ran to seed and produced skimming dishes which were unsuitable in the large classes. We need not consider this form of section any further because it is of no use in the small cruiser class; there is no room inside. We have, by a consideration of these simple facts, come to certain useful conclusions, one might almost call them principles.

We cannot alter the total weight of our yacht, which is one of the two factors producing stability, but we can alter the lever S, and in two ways: we can adopt a mid-section that, while affording the necessary inside room, throws the centre of buoyancy as far outwards as possible for a given angle of heel; or can lower the centre of gravity by reducing the weight of the hull, gear and equipment, to the lowest figure consistent with adequate strength; and this reduction is essential in the spars and gear. Construction must be scientific rather than weighty. Diagonal bracing, good workmanship and the choice of light material will produce a hull that is just as strong as the fishing boat type of building. We build our yachts to sail and not to break ice.

What about our freeboard and the proportion of length to beam? In small and medium size cruisers it is quite safe to make the least freeboard one-tenth of the L.W.L. In most cases this will mean that an inclined water-line drawn from the rail amidships through the centre of the L.W.L. will show an angle of 25 degrees. So we may say that for a yacht whose beam is not less than one-third of the L.W.L. we can decide upon the freeboard by drawing an inclined water-line at 25 degrees. It will be seen in Fig. 3 that this angle in *Rory* is 30 degrees. But *Rory*'s beam is considerably less than one-third of the L.W.L.

The relation of the beam to the L.W.L. can be determined by the formula $\sqrt[2]{\text{beam}} = \sqrt[3]{\text{L.W.L.}}$ Thus: the square of 2 is 4, and the cube is 8. In other words an 8-ft. sailing dinghy may have a beam of 4 ft., and so it can have. The square of 3 is 9 and the cube 27. A load water-line of 27 has a beam of 9. Similarly, taking 4 as the basic figure, we have a beam of 16 associated with a L.W.L. of 64. These figures are of course approximate but the rule which was given to me by Mr. G. Dunn is a most useful guide. Later on I shall give a rule* to determine the proper amount of sail area.

When, after a consideration of what has been said, we have determined the main features of the yacht, and the mid-section, we must decide how we are to fill in the drawing for the bow and the stern. We can think of our hull as a cellophane bag bounded amidships by the chosen mid-section, and filled with thin putty. We can massage the contents fore and aft in any way that we like. If we massage the insides for'ard we have the old-fashioned cod's head bow, or we can squeeze the putty aft, and obtain the square quarters and fine bow that has for decades ruined the performance of our cruising yachts and fishing craft. Both bad, but the latter much the worst. But why not strike an average and make the stern match the bow? This can be done with certainty by the methods described by Rear-Admiral Alfred Turner, which will be described in a later chapter.

Having now considered the rough outline of our yacht both as regards her outward shape, her proportions, and the disposition of her displacement, we are ready to begin our actual work on the drawing-board. The subject I have chosen is *Sinah*, a yacht that I designed some few years ago. She has been built in Denmark and has proved to be a fine sea-boat, very handy and perfectly balanced on her helm. She is, I may say, the most difficult type to design, and I most strongly advise the beginner to choose an easier type. He should first take a design from one of the yachting papers; enlarge it with proportional compasses and then fair it up. Then he should choose for his first original work a hull with slack bilges and with great length in proportion to her beam; say a 100-ton schooner. The long sweeping water-lines and diagonals are far easier to draw and to ink in, than are the sharp curves found in a small yacht which must have a large amount of beam.

* See page 74 *post*.

A professional designer makes all his drawings on tracing paper, and does not ink them in, but at once makes a tracing on tracing cloth. Whereas the beginner will have to make hosts of corrections he will be wiser to use Whatman. I prefer to make my *lines drawing* on paper, but cabin plans, construction plans and sail plans I now make on detail paper. Erasures are made with a sharp knife and finished off with an ink eraser. It is wonderful what can be done in this way. Once I had completely finished a drawing and then upset the ink bottle over it. An hour or two with a pen-knife made all good.

CHAPTER II

The Drawing

AFTER these general considerations we are now ready to commence the actual drawing. I am assuming that the preliminary sketch has been made, and that the main dimensions have been chosen. Although I think that the beginner will be well advised to begin the design at half the size of the final drawing, I am leaving out this stage and going straight to the full size plan.

It would naturally be more methodical were I to develop the design of *Sprite-of-Arden*, but she has not so far got beyond the preliminary sketch stage. When some three years ago I designed *Sinah* I had in mind that at some time or other I should write an article on yacht designing. Accordingly when I had completed the design in pencil I traced only as much as is shown in Fig. 6 and had a true-to-scale reproduction made of this tracing. Then I traced some more, as shown in Fig. 9, and again had a true-to-scale print made. Finally I completed the tracing and the final stage will be seen later on.

Two kinds of reproductions are in common use. The "blue-print" is well known. It is cheap, and also nasty. The prepared paper is exposed under the tracing and then developed by a wet process. I think that the basis of the paper is potassium ferro-cyanide, and that it is developed by immersion in water, but it may be more complicated than this. The drawback is that the paper shrinks in drying and the dimensions are inaccurate. The true-to-scale print is perfectly accurate because it is made by a dry process. It is generally in black and white and exactly resembles the original drawing. There is one thing that must be noted; an indifferent tracing will give a reasonable blue-print, but to obtain a good true-to-scale the tracing must be good with sharply defined lines in black ink. Red lines, not magenta, will appear definitely but faintly in a blue print, but very badly on a true-to-scale. This is a digression, but perhaps a useful one. It would be most interesting if someone who understands these things would write an article upon tracing and reproduction. It has never been done and I am sure that we should all welcome it.

We now pin our sheet of Whatman on to the drawing-board with four drawing-pins or attach it with Durex drafting tape. I personally have not so far used this tape, but it is probably better than pins. Get the best pins. The kind with a folding handle attached will save much trouble and perhaps damage to finger nails. If you cannot obtain this type get a small instrument shaped like a two-pronged fork for removing the drawing-pins from the board. See that they have steel shafts; those with shafts of the same material as the heads are a constant irritation, for the shafts ever bend flat. When the paper has been carefully fixed down and all bulges have been smoothed out, take the large T-square and rule in the load water-line, and the centre line of the half-breadth plan. The body plan must be adjusted so that it fits into a convenient place on the board, or it may be drawn on a separate sheet of paper on the small drawing-board. This is the better plan for the beginner because it makes inking-in much easier. Some place the body plan on the sheer plan as shown in the drawing of *Sprite-of-Arden* (Fig. 1). This plan makes for accuracy, because any inequality in the spacing of the water-lines is the same for both plans and does not influence the design. But it does not look nice, and may confuse the beginner. I call it a "messy" way of doing things. Reference to the original one-quarter scale drawing will show the correct distance between the sheer and half-breadth plans, and if a rough scale drawing of the mid-section be made on paper and cut out with scissors it can be adjusted between the two drawings as soon as the sheer and deck plans have been pencilled in. These little arrangements are necessary, otherwise one may find the two plans too close to each other, or the sheer may come too close to the top edge of the paper. We have before this chosen a suitable scale. Our sheet of Elephant Whatman measures 28 in. by 32 in. *Sinah* we find from our original sketch to be 31 ft. long, so it is obvious that the scale cannot be an inch to a foot. We choose a scale of $\frac{3}{4}$ in. to 1 ft. The yacht will now measure $23\frac{1}{4}$ in. long and will fit nicely upon our paper. A professional designer would generally use a scale of an inch to a foot for *Sinah*, but this means a large board, one that does not suit the modern bijou residence, but asks for a drawing office. A large drawing means a lot of space to cover and the designer must have a proper desk and stand to his work.

Draw the L.W.L. and the centre line of the half-breadth plan, then a vertical line which will represent the centre of the load water-line. Whereas the forward overhang in *Sinah* is longer than the after overhang, and because we have to find room for the body plan, this centre line will be somewhat to the left of the centre of the paper. Here I may remark that it has become a convention to draw yachts with the bow to the

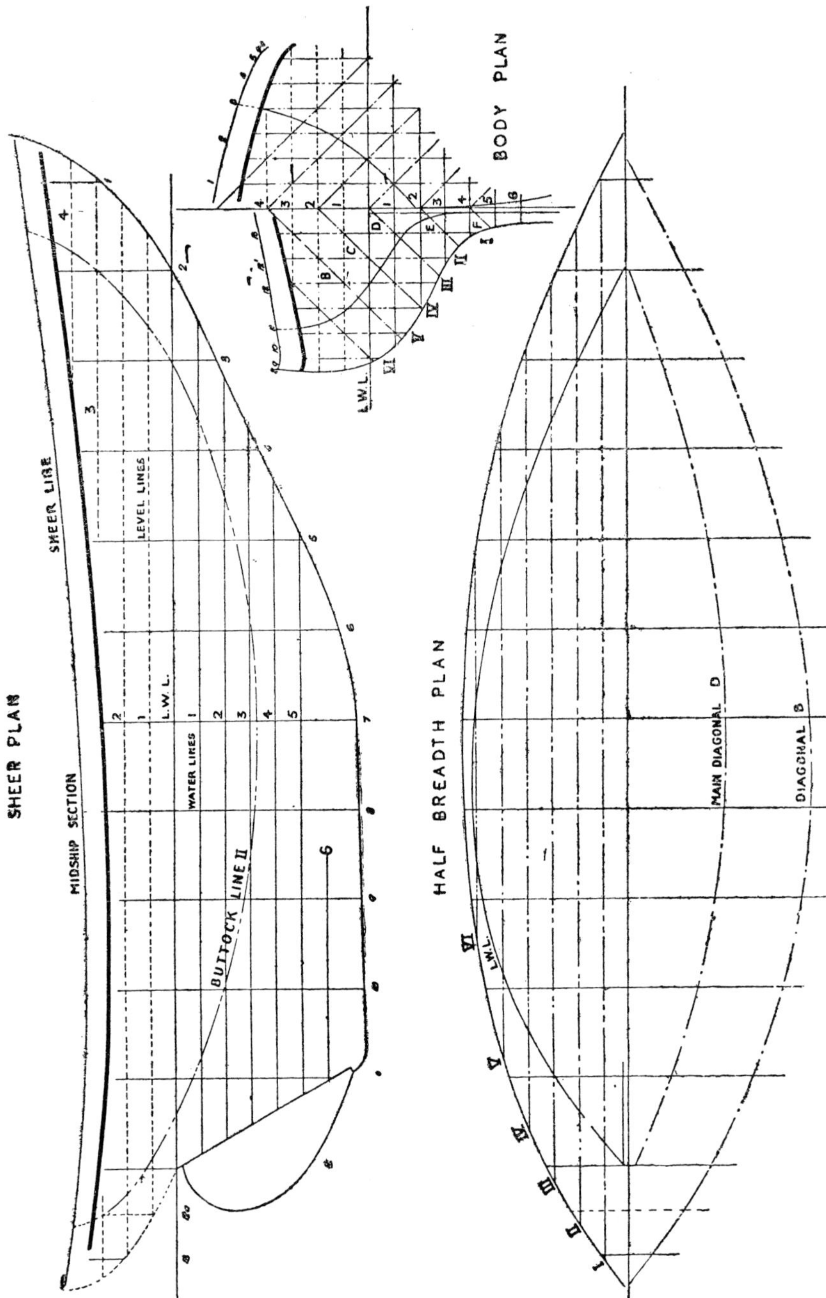

Fig. 6. *SINAH.* The drawing has been started and here can be seen the mid-section, the key sections, the load water-line one bow and buttock line and two diagonals.

right. Only one or two designers have worked in reverse fashion; conspicuous among these was the late Mr. Clayton. This centre line must be taken right across the board so that it cuts both the L.W.L. and the central line of the half-breadth plan.

The L.W.L. of *Sinah* measures 24 ft. Take the $\frac{3}{4}$-in. scale and measure 12 ft. forward and 12 ft. backward. Make a dot with the HH pencil. These two spots represent the forward and after ends of the L.W.L. I myself in common with some professional designers always divide the L.W.L. into ten parts. This makes the calculations simple, and if all one's water-lines are decimally divisible it is easier to compare two designs than if the two have their sections spaced in different proportions. Some designers do not divide the L.W.L. into equal parts at all, but leave a bit over at each end. I can see no reason for this procedure; it makes the calculations very difficult, but may be helpful in some cases in measuring the distances between the frames or moulds. From this point of view it would be quite legitimate, in fact better, to divide this particular L.W.L. into twelve parts for then each would be 2 ft. apart, a very convenient measurement for the spacing of the moulds. As it is they are 2.4 ft. apart, a nasty fraction for the builder. This measurement must be made for the half-breadth plan as well. It will not do to square down with the T-square. This method is inaccurate and absolute accuracy is called for. Now divide the L.W.L. into ten parts. You will have a scale divided into tenths. Space out 2.4 ft. with your pencil compass and make the necessary marks on the L.W.L. Now place the scale close up to the L.W.L. and check the divisions till they are dead accurate. Any inaccuracy will give rise to untold trouble later on. Now with the T-square join up the points on the L.W.L. and the centre line of the half-breadth plan. Of course if these is no rock on the T-square, if the square is a perfect right-angle, and if the drawing-board is true, then these lines will square up with the T-square both from the top and bottom side of the board. In my practice they never do so, and therefore it is necessary to measure off the divisions above and below and join up.

We have now a L.W.L. and eleven vertical sections. Now take your small sketch and measure off the sheer line with compasses, either stepping each measurement off four times with ordinary compasses or once with proportional compasses. These are not really essential but they are useful if one wishes to alter the size of a design. Pencil in the sheer line with a stiff batten and four or five lead weights.

The next item is the stern post. In this case it is inclined 60 degrees to the L.W.L. With the 60-degree set-square or with a protractor lay off the stern post at the required inclination. The underwater body can now be pencilled in from measurements taken from the preliminary sketch. This will include the bow overhang and in this case the counter stern. I think that now it is a good thing to ink in the drawing as far as it has gone, including the rudder. This will be found to correspond with one of Dixon Kemp's "pears." Do not ink in the canoe-stern. When the design is approaching completion it may be found that it does not quite flow naturally into this stern, and slight modifications may be called for to make an harmonious end to the hull. The curve should be parabolic, but the length may need modification. In Fig. 6 the stern line is shown by a pecked line which means that it has not yet been inked in.

We now have our side elevation of the hull. The next stage is to put in the water-lines and the level lines. The horizontal lines above the L.W.L. are often wrongly called water-lines. They are level lines. Similarly it is sloppy to talk of buttock lines when speaking of the whole curve from stem to stern. It is correct to name the bow portion a bow line, and the stern portion a buttock line. Together they ought to be spoken of as bow and buttock lines. The water and level lines can be spaced at any distance. In the case of *Sinah* I have spaced them 8 in. The simplest way to measure them is to take the scale of an inch to a foot and measure them by this scale to half a foot; 6 in. multiplied by $4/3 = 8$ in. See to it that these lines are perfectly accurately spaced by measuring at each end. Do not trust the T-square with the water-lines which must now be inked in. It is essential to ink in the construction lines, otherwise they will be rubbed out when the various sections are modified.

Now come down to the half-breadth drawing. It is already divided up into sections. Taking the measurement from the initial sketch, mark off the greatest breadth at the deck line on section 8, and with the T-square drop down the forward and aft ends of the yacht. Correct the length of the fore and aft overhang by measurement, for, as we have stated, the T-square is inaccurate. We now have three points for our deck line. That on the mid-section, and the front and back end. With a batten and five weights run in the deck line, taking the measurements from the preliminary quarter-scale sketch. We now locate our body-plan on the board. I told you that the easiest plan was to make a template of the mid-section and with it discover the best position. You may go a long way with the body-plan and then find that the bottom of the keel runs into the deck line of the half-breadth plan. It does not matter, but looks bad.

We now leave the sheer-plan and half-breadth plan and divert our attention to the body-plan. Rule in the centre line, the load water-line and the water-lines and level lines exactly as in the sheer-plan. This stage marks the actual beginning of the design where the initial sketch left off. The mid-section alone gave a clue to the fundamental form of the yacht. Transfer this mid-section to the present drawing, just in the same way as we transferred the sheer, preferably with proportional compasses. Before, however, we do this we must add some water-lines to our initial sketch. They were not included in Fig. 1. These will be 2 in. apart using a scale of $\frac{3}{16}$ in. to 1 ft. or 4 in. if a $\frac{3}{8}$in. scale is chosen. It is best to get some box-wood scales. I use an inch to the foot divided into tenths and also inches, one on each side. Then another $\frac{3}{4}$ in. to 1 ft. and $\frac{3}{8}$ in. both divided into inches. A third has a $\frac{3}{8}$in. scale on one side divided into tenths, and on the other side one of $\frac{1}{3}$ in. to 1 ft. divided into tenths. I find this very useful in drawing curves of areas. A metric scale may be handy from time to time.

Two more sections only are now necessary to fix the character of the hull. They are what I call the fore and aft key sections. I strike them at a spot one-tenth of the L.W.L. from the fore and aft ends of the L.W.L. This is purely an empirical choice of position; the key sections must be somewhere in these situations, and we already have a division line at one-tenth from the ends of the L.W.L.

How are we to draw these two curves? I can best answer this question by detailing how I choose the curves shown in my drawing. I had for many years been trying to design a balanced hull; one that, like a model, would sail itself to windward unattended and would be finger-light on the tiller off the wind. I knew that hulls designed to fit the metacentric theory had these qualities, but I could not quite understand the details of the method. I studied these vessels, among them Robert Clark's *Mystery*, and I noted that the bow curves were fuller than it was usual to make them with reference to the stern sections. So I drew section 3 with a good bold curve. It is actually a segment of a circle. Its lowest point was determined by a measurement from the sheer-plan from the water-line level to the keel level on section 3. The top end of the curve can be localized by two measurements. The height of the sheer line at section 3 and the width on the deck line taken from the tentative deck line on the half-breadth plan. The after key section, section 11, was drawn in the same way. These were my idea of what the curves ought to be. At present both must be regarded as purely provisional.

What sort of a load water-line will they give? We have now five fixed points to locate the L.W.L. The fore and aft ends have already been

PLATE 1

(above) *Memory* (18′ LWL). Built 1912 by E.L. Woods, Cantley, for H.J. Suffling. The first HB boat known to have been built in Britain. For many years T.H.B.'s favourite design. She is now in Eire.

(left) *Seagull* (18′ LWL). Built 1913 by E.L. Woods for N.R. Suffling. Now renamed *Fleetwing*, she is still East Coast-based.

PLATE 2

(left) *Argo* (21.8′ LWL). Built 1914 by Burt & Son, Falmouth, for W.A. Bartram. Falmouth-based until recently and now ashore for a refit.

(below) *Sandook* (25′ LWL). Built 1897. A Plymouth Hooker designed/built by Burlace, Plymouth. Owned by THB from 1912–1934, with no engine for nineteen years. *Sandook* with second jib and double-reefed main demonstrates the shortage of luff when a gaff sail is hard-reefed.

fixed. We have the width at the mid-section, and the width at each key section. Mark these three measurements on the half-breadth plan, and with a batten and not more than five weights run in a water-line through these five points. Let us assume that it looks satisfactory, a nice sweeping curve fore and aft, somewhat blunter aft than fore. We shall now apply two more tests to the design, for it has now almost become such. We have ruled in the level and water-lines across the body-plan. Continue by adding a similar series of lines spaced at 8 in. but placed vertically. These are the bow and buttock lines projected in a vertical plane. The whole body-plan is now covered with a series of squares. In inking in it is usual to use a continuous line for the water-lines and pecked lines for the level lines. This is merely a convention. Now take the T-square and a 45-degree set-square and put in the diagonals. I have started them from alternate water-lines, giving six diagonals. It is quite obvious without consulting Euclid that if the work is accurate each diagonal will cut the intersections of the vertical and horizontal sections. I make it a practice to locate my diagonals at 45 degrees, but the majority of designers adopt various angles. I am in good company with Laurent Giles. I now propose to run in a diagonal and a bow and buttock line. The top diagonal but one, labelled B, Fig. 6, is most suitable for our purpose. We have five fixed points to locate this curve. It starts fore and aft where level line 3 cuts the stem and the canoe-stern profile line. These two points must be squared down to the half-breadth plan and corrected by measurement. Its mid-point is ascertained by measuring the length of the line on the body-plan from where level line 4 cuts the central axis of the body-plan to the line of the mid-section. This measurement is transferred to the half-breadth plan on its lower side opposite to that used for the water-lines. A similar measurement to sections 3 and 11 gives the two intermediate loci. Again we take the spline and five weights and sweep in a curve through the five points. If this gives a nice sweet curve we have corroboration that our key sections are good.

Finally we want a bow and buttock line. I have chosen one 16 in. from the mid-line, labelled II on the body-plan. Before we can do this we must rule the bow and buttock lines on the half-breadth plan. They are seen as six lines parallel to the central line and spaced 8 in. from this line and each other. We have not now so much latitude in shaping our curve for we are limited by seven loci. The two ends of the curve are found where the buttock line on the half-breadth plan cuts the deck line, measured from the nearest section, in this case 2 and 12A. Two more loci are where the same line cuts the L.W.L. fore and aft. The last three are obtained from the body-plan by measuring down from the water-line to the intersection of sections 3 and 11, and the mid-section with the vertical buttock line II. That is to say that a vertical plane passing through the hull at a distance of 16 in. from the mid-vertical plane will have the shape shown on the sheer-plan. It is located by its intersections with the deck fore and aft, the L.W.L. fore and aft, and with the mid-section and the key sections fore and aft. The central part of this line can be swept in with a spline, but the fore and aft curves will be drawn with the aid of appropriate curves. If this bow and buttock curve is an harmonious one we can be pretty certain that our design is shaping well.

The next stage will be to plot in the sheer line on the body-plan. At present we have only three sections to deal with: the mid-section and sections 3 and 11. First measure the heights of these three sections on the sheer-plan and transfer these measurements to the central axis of the body-plan, and from the spots found rule three horizontal lines representing the heights of the deck line at the three sections: mid-section, 3 and 11. Now from the half-breadth plan measure the widths of the deck at these three sections. Mark them off on the appropriate horizontal lines already ruled in. A line drawn through these points will give the sheer line projected on to the body-plan. Finally we put in the main diagonal, starting at the fore and aft end of the L.W.L. We measure the distance along diagonal III from the axis of the body-plan to the mid-section and to sections 3 and 11, and we have five points to locate the curve of the half-breadth line. If this curve is a fair and harmonious one and can be drawn in with not more than five weights on the spline, then we can assume that our design is fair as far as it goes.

It is now a complete skeleton of the final design and if in its present condition it were given to three designers to complete, the final result would be practically the same in each case. The mid-section, the key sections, the deck line, the L.W.L., the bow and buttock line, and the two diagonals, give a series of fixed points for any number of sections and stereotype the shape of the hull. It is essential that all the measurements made so far shall have been accurate. A *nearly right* will ultimately grow into an *entirely wrong*, and enormous trouble and copious alterations will be caused by a little initial inaccuracy; I was thinking of saying laziness!

We have now reached a critical stage in our work, and it is a good time to review it before it goes so far that modifications would need a completely new drawing. At the present time we can alter the character of the hull, or remedy defects, merely by rubbing out a few pencil lines. Are we satisfied with the sheer of the yacht, or with its general profile?

Is the stern right? Perhaps not. I think that were I designing *Sinah* again I should cut out some of the deadwood aft. The stern post could come say 6 in. further forward, and more slope could be introduced into the stern post. To balance this alteration it would probably be necessary to introduce a hollow into the forefoot with other alterations to maintain hull balance. Some may say "why cut away the after deadwood? It conduces to good running."

The answer to this is a direct negative. It is to improve the running properties of the hull that I wish to cut it away, and incidentally by reducing hull surface to improve the speed in light winds. Experiments with models have shown most conclusively that the long keel will not run, and that the water from the garboards must be able to get cleanly away from the hull, and not get confined under the run, inducing pressure there and erratic running. The pure fin keel runs best. This is heresy to the old shell-backs, but it is fact, always assuming that the hull is balanced.

But before we go further it is most essential to find out if our hull is balanced. I know of only one method to discover this, the metacentric theory of Admiral Turner. Being a new thing it was bound to meet with severe opposition. I care little for the theory, but I know that all my metacentric designs have produced hulls that run straight. One of our best-known designers who at first would have none of it, now tells me that he has criticized and analysed it in every way, including a mathematical examination, and that it is correct. In *The Yachting World*, Vol. XCII, page 123, I read: "The hull of *Fidalga I* was designed by Mr. K. C. Barnaby (Chief Naval Architect to the Southampton yard of Thornycroft's) for efficiency under sail and was balanced up by the metacentric system." I think that if we know that in practice the system produces a balanced hull, and that an architect of the calibre of Mr. Barnaby employs it for the design for his own use, not I may say of a pure sailing craft, but of a "motor-sailer" or "fifty-fifty," we should employ it for our present purpose.

It is not possible to make a complete analysis at this stage, but we can do enough to show us that we are on the right path to correct balance. I intend to use Mr. Welch's simple procedure on the body-plan, Fig. 6. To make the matter easier of comprehension I shall now reduplicate the key sections and the mid-section in pecked lines. You can if you like, and it is a good plan, draw two body-plans, one a complete bow, right and left, and the other a complete stern. Now put in the inclined water-lines passing through the middle of the L.W.L. on the body-plan to the top of the deck line at the point of lowest freeboard, see Fig. 9. That is to say we heel the ship till her deck is level with the water. In this type of hull, as I have already said, this line forms an angle of 25 degrees with the L.W.L. As we have tidy minds, we now put in the two inclined water-lines at this angle. Finally draw two lines at right-angles to these inclined water-lines starting from the middle of the L.W.L. These lines are the Welch axis, not the metacentric axis but one parallel to it.

Take some good detail paper and trace out from Fig. 6 the three inclined sections: mid-section 8, key section 3 and key section 11. On each paper section rule in the Welch axis, and poise each on a razor edge as in Fig. 7*. The three inclined sections are seen in Fig. 7a. Having found the line of poise rule it in on the section parallel to the Welch axis. Lay off the distance between the Welch axis and the poised line on the half-breadth plan along each appropriate section, that is section 3, mid-section and 11, and then strike a curve through the five points: the fore end of the L.W.L., the three section poise marks, and the after end of the rudder projected down to the mid-line of the half-breadth plan. This line is the probable "shelf" of the hull, as far as the present data enable us to plot it. There are three types of good shelves, and two bad ones.

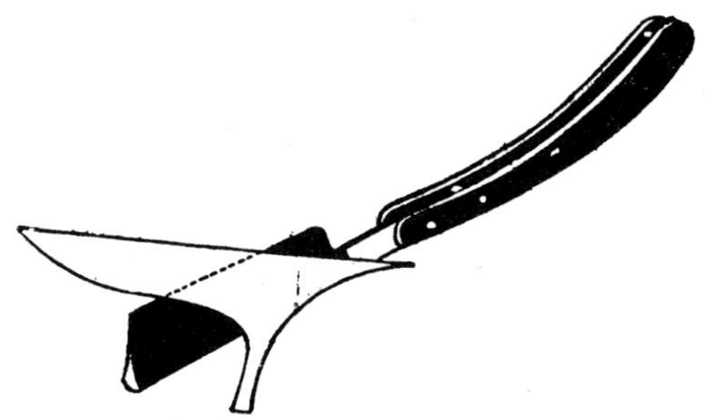

Fig. 7. Poising a section on a razor edge.

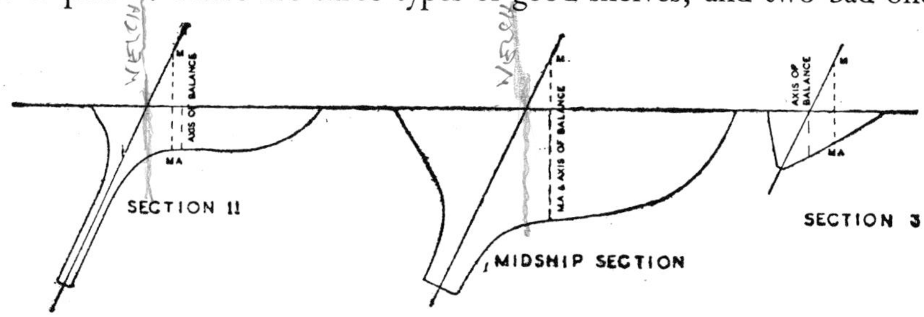

Fig. 7a. Inclined sections.†

In the first place the shelf may be straight and parallel to the axis, the best form, but one difficult to obtain except in shallow draught yachts. Then it may form a symmetrical curve, concave or convex according to the type of hull, again a good shelf. Finally it may be asymmetrical, approaching

* NOTE.—When poising, the razor edge must be kept parallel with the Welch axis.

† M.A.—metacentric axis, see page 108 *post*.

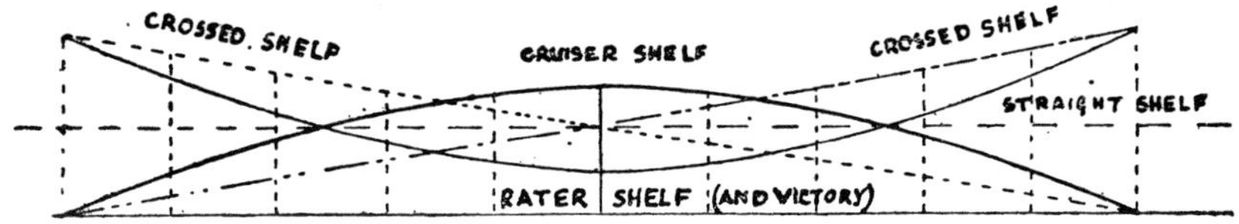

Fig. 8. Types of metacentric shelf.

the axis at one end and leaving it at the other. This is the "crossed shelf" found in our fishing boats, Brixham trawlers, Falmouth Quay punts, and most of the older cruising yachts. It is not present in Scandinavian boats, Arab dhows, Chinese junks, and such like. These types of shelf are figured in Fig. 8. The shelves are referred to the Welch axis, first described by the owner and designer of that well balanced and highly successful yacht *Fidelis*. Later on, the shelves, or rather the actual shelf of a yacht, will be plotted against a real metacentric axis. The probationary shelf found by the method described has been plotted in below the central axis of the half-breadth plan in Fig. 10. It is obvious that it is a good one, a clean symmetrical curve. We may rest assured that the skeleton we have constructed will work up into a well balanced hull that will "hold the road," and not tire out her steersman. A comparison of the inclined sections 3 and 11 conveys the impression that the centre of gravity of 11 would be further to leeward than that of 3, but one must not forget that this section has a long tail of deadwood and so the two sections balance. Were this tail cut down as shown by the dotted-in profile in the sheer-plan in Fig. 33,* then the poise would be different, and this might make it necessary to fill out the bow sections to maintain equilibrium. One would not naturally think that cutting down after deadwood would call for a fuller bow, but so it is, and without the metacentric system one would not be aware of this necessity.

We have now completed our study of the skeleton and are in a position to carry out the donkey work of filling in the rest of the sections and fairing up.

I have been asked why I so constantly say that the spline must be kept in place by five weights. The reason is that any curve that calls for a larger number has no right to be in the water-lines or the diagonals of a yacht of the ordinary type.

* See page 131 *post*.

CHAPTER III

Fairing Up

We have now completed the skeleton of our yacht. We have made a preliminary investigation of its balance, and have considered any alterations which might improve it. Let us assume that we are satisfied. The next stage is to put flesh on to the skeleton. This we may do in a variety of ways. We can introduce all the vertical sections on to the body-plan and then proceed to check them up with water-lines, diagonals, and bow and buttock lines, or we can introduce two more sections and then check them in the same way. I think that it will be best to fill in all the vertical sections. For each one we have several fixed points and it is necessary only to join up these points freehand to fashion the section. For each section we have first of all the width on the L.W.L.; then we have the width at the deck line, and the draught of the section; there are two intersections with the two diagonals, and one with the bow and buttock line.

In Fig. 9 these sections have been introduced into the body-plan, but it will be noted that the after sections have not been completed down to the bottom. The reason for this will be explained later, but it is obvious that at the moment we have no guide to the form of the curves in this position. It is possible but wildly improbable that the curves we have now inserted are final and correct, and that the design as far as the body-plan goes is finished.

The next stage is to fair up the sections, and I fear that the beginner will find this process tedious and time consuming. The keynote is accuracy and sense of a true curve. Never let anything pass that is not perfect. Do not distort a curve, however slightly, to make things match. This trifling error will gather force and increase, till finally one curve will be found to show so much distortion that it is obvious that something is very wrong, and vast alterations may be called for. When faced with a difficulty, and they will occur mainly at the after end of the hull, do not panic. You find that your original bow and buttock line will not conform to the new sections in one or two spots. Do not at once make a large alteration, but recollect that each modification affects four curves, the vertical section, the water-line, the buttock line, and the diagonal.

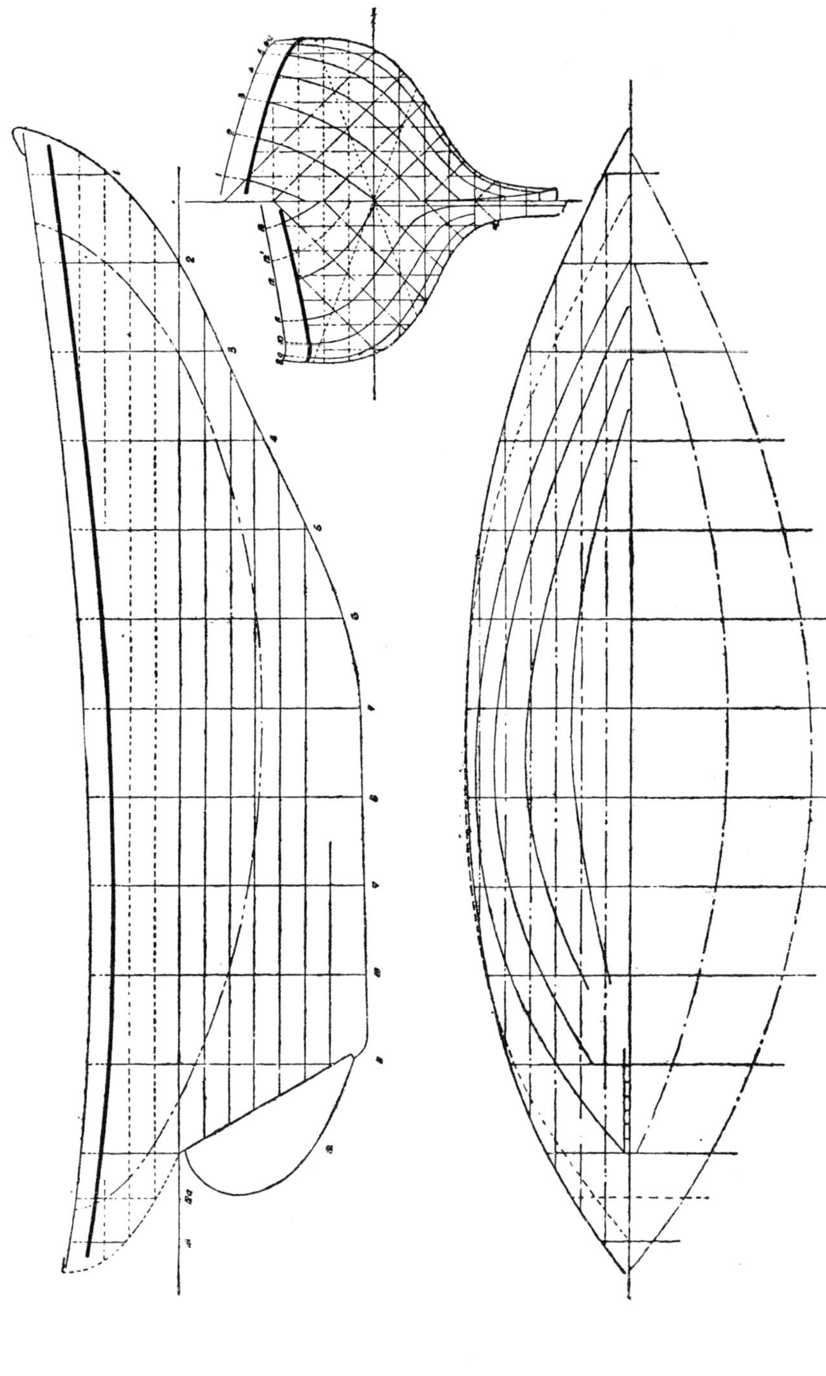

Fig. 9. *SINAH*. The design progresses; most of the sections have been filled in on the body plan, and the level lines on the half-breadth plan.

A trifling alteration to all or some of these will prevent a larger modification of one or two of them.

The construction of the hull must not be lost sight of. It looks very pretty to bring the water-lines to a point forward and aft, but in reality the stem of a yacht has not a razor edge, and there is such a thing as a stern post aft. Decide at once upon the thickness of these members. For *Sinah* the stem can be half an inch widening gradually as we pass down the keel. The stern post is 4 in. thick. Lay off these measurements on the half-breadth plan. This has been done in Fig. 10 for the stern post. This is crossed by a series of transverse lines which should be equally spaced. These represent the loci where each water-line ends aft, and are obviously obtained from the sheer-plan by measuring from section 11 along each water-line to the stern post. Those who have a pathetic trust in T-squares may square down. Extreme accuracy here does not matter. Talking of T-squares I have just discovered that my small one is not an exact right-angle! It has not led me astray because I have never trusted it, but I have used it for inking in the sections of the sheer-plan in many of my designs and have wondered why the section did not carry down sweetly to that of the half-breadth plan. If you are working on a half-scale drawing you can fair up the final full-scale drawing by a method varied from the one employed for the first drawing. For example I propose now to fill in the water-lines and one level line, and then to put in all the diagonals, and continue with the bow and buttock lines, completing the design with the remainder of the level lines. In the final drawing, having put in all the vertical sections, I would begin to fair on bow and buttock lines alone. The combination of two methods insures complete fairness. Why all this struggle for absolute fairness? In the first place if the yacht be built by a good builder who properly lays out his work, the loftsman will not be at all complimentary to the designer who has given him so much trouble to get the lines fair, and when he has faired them the result may not be exactly what the designer intended. If on the other hand the builder is one of those who just takes the sections from the drawing, enlarges them and makes his moulds from each individual section, then the resulting yacht will perhaps be what the designer has drawn but not what he wished to draw. Finally, if the hull were built in steel and were not carefully laid off, then the resulting yacht would be all flats and cockles. Wood construction is kind to bad designing and lofting because it tends of itself to take fair curves. In fairing up the sections it is I think best to begin at the L.W.L. and then fair up and down. The water-lines that have been introduced will show that the underwater part of the hull is now reasonably fair, and the level line shown in the half-breadth plan as a pecked line will give fairness to the top-sides. The reverse curves of the run have not yet been introduced because we need the inner bow and buttock line to find out their best shape. We now leave Fig. 9 and pass on to Fig. 10 which shows the completed design. It will be noted that on account of the tumble-home of the top-sides the level line intersects the deck line. At a slightly later stage when we are quite sure that the assumed shape for the canoe-stern is correct, we shall ink in the deck line, and then check over the level line. At the moment our two pencil lines are apt to become confused.

The next stage is to insert the inner bow and buttock line, No. 1 in Fig. 6. This will enable us to design the run of the ship, thus filling in the reverse curves of the water-lines, and the lower terminations of the vertical sections in the body-plan. If we now run in the two lowest water-lines, E and F, we shall have completed the keel part of the design, unless we find later on that some modification is necessary to bring the lead keel to its right dimensions. I would now put in all the diagonals, being careful that the spline is not constrained from a true and harmonious curve. Small alterations will be necessary to the sections and the water-lines and perhaps even to the two buttock lines. If the work has been faithfully done the remainder of the buttock lines ought to run in quite sweetly with minimum alterations. They have to make a large number of intersections with the water-lines and the sections.

It will be noted that the top diagonal in Fig. 10 ends blindly at its after end. This was because I did not know the correct position of its ending on the mid-line of the half-breadth plan. Since making this drawing of *Sinah* I discovered a diagram (Fig. 11) and explanation by Mr. L'Estrange Ewen in *The Yachtsman* dated 17th January, 1901 (Vol. XXI, page 35). I quote from his accompanying letter: "In fairing up the lines of a yacht it is sometimes found that in placing the diagonal to cross the sections in the body-plan as near to an angle of 90 degrees as practicable, the result is to cut the centre line above the level of the deck. To find the forward ending of diagonal *a-b*: find where the rail half-breadth equals *g-f* in the body-plan as at g_1f_1 in the half-breadth plan. Draw $g_1f_1f_2$ and set off g_1f_2 equal to *a-f*; f_2 is the ending of the diagonal. To find the after ending of the diagonal *a-e*: if the diagonal fall clear of the knuckle in counter, proceed as for fore-body, but if it cuts the knucklc, as *c-d*, draw a section through k_1h_2. Set off k_1k_3 in half-breadth equal to *c-k*; *k*-k_3 is the termination required." In a later number of *The Yachtsman*, 14th February, 1901, W. P. Stephens criticized this method

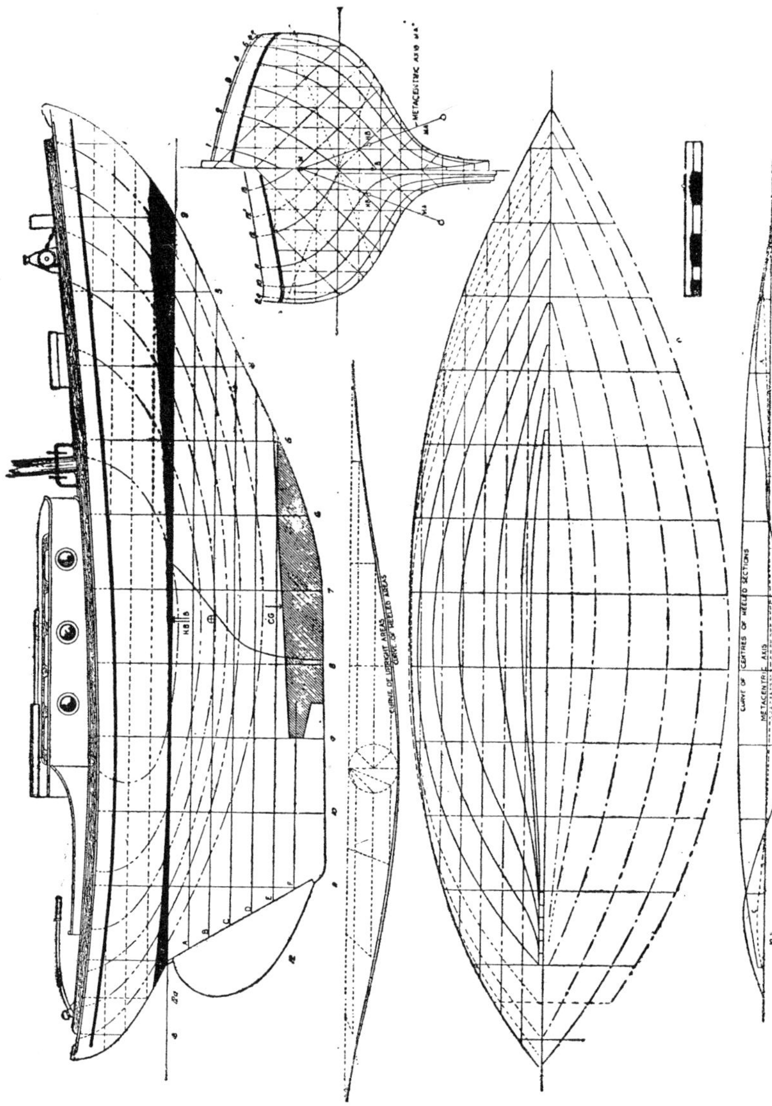

Fig. 10.
SINAH.
The completed design.

T.M. 8 tons.
L.O.A. 31 ft.
L.W.L. 24 ft.
Beam 8 ft. 6 in.
Draught 5 ft.
Displacement 6 tons.

of dealing with the knuckle, but L'Estrange Ewen replied that his method was for all practical purposes sufficient.

The design is now nearly complete, but the most important part, the top-side, claims our attention.

Carefully check over the dimensions of the deck line. During the fairing some alterations may have been made and you have perhaps forgotten to carry them up to the deck line. In fairing, as I have already pointed out, any alteration in a line may involve three others. You modify a buttock line; this calls for a corresponding change in a water-line, in a diagonal and a vertical section. It is fatally easy to forget one of them, generally a diagonal. Having satisfied yourself that the deck line is correct run a spline round it, and line it in with a sharp HB pencil. Now take off the spline and have a good look at it, and see that this spline-line registers with your original line. If not make the necessary alterations to the body-plan. Such a modification does not entail any further alteration, except perhaps to one of the upper ends of the bow and buttock lines. When all is ship-shape, carefully wipe the paper clear of any bits of rubber and ink in the line. See that the noses of the weights do not project, but if at the last moment you find that the pen is going to touch one, leave out this section of the line and ink it in afterwards. Do not tilt the pen or the ink may touch the spline, and then surface tension will do the rest; capillary attraction will draw the ink under the spline and there is a nice mess. Do not panic; leave the spline *in situ* till the ink is dry, and then get to work carefully with a sharp pen-knife—you will have a small stone to hone your pen-knife, which needs doing frequently—and an ink eraser. When you ink in the gap in the line see that there is very little ink in the pen, for it is liable to spread slightly on the scratched surface and make a fuzzy line. I was advised to use a disused razor blade for erasing, but it did not, at any rate in my hands, work well. Perhaps a practical draughtsman, who I hope will write a paper on tracing and reproduction, will also deal with erasures. It is an important subject, for unless you are exceptionally clever you are certain to be faced with blots and blemishes.

I have just obtained a tube of a substance called "Erazol." This is an abrasive mixture which can be squeezed upon the blot or false line, and then with a clean cloth the Indian ink can be gradually dissolved by light rubbing. The method leaves a clean smooth surface upon which the lines can be redrawn without running. It will also work on tracing cloth and on paper, but more time is necessary to complete the erasure.

We are now ready to finish the top-side. I would rub out all the

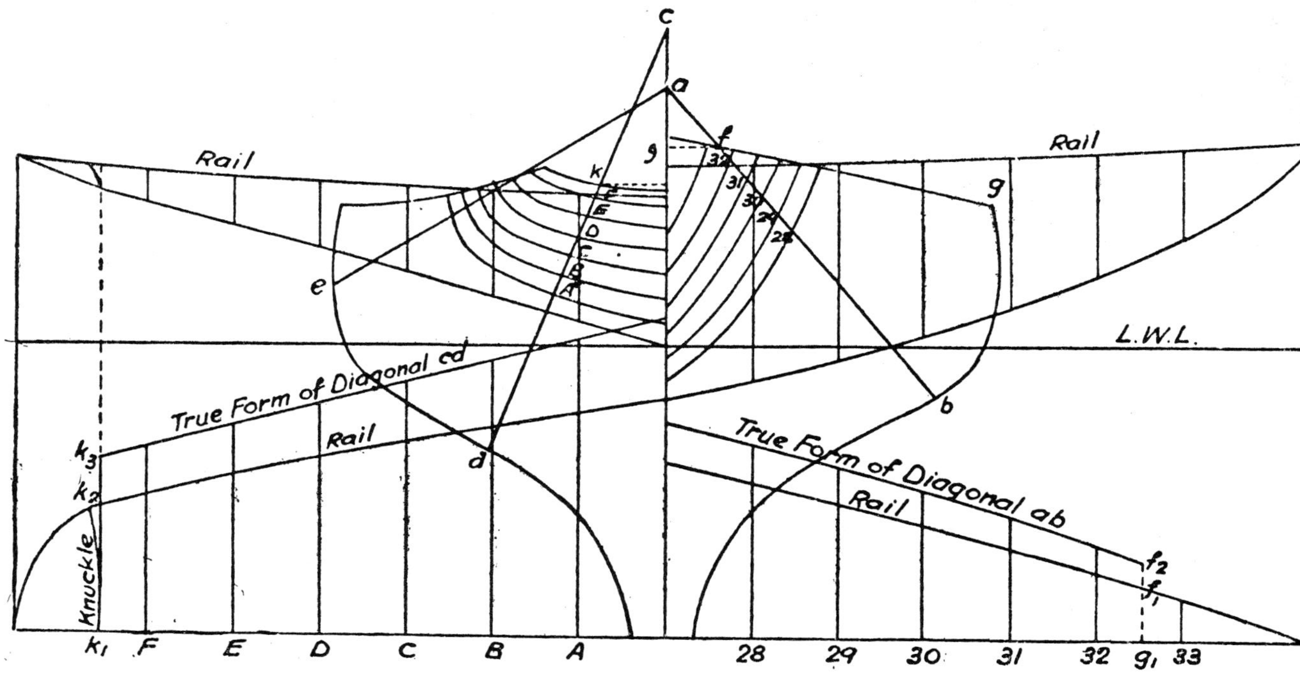

Fig. 11. L'Estrange Ewen's diagram for finding the correct endings of diagonals. (From *The Yachtsman*, 17th January, 1901.)

original lines in the central part of the vessel, and get it nice and clean. Now begin with level line 2 in Fig. 6, measure the breadths or rather the half-breadths from the body-plan and transfer the measurements to the half-breadth plan. I have not, I think, mentioned how these should be made. I believe that the proper way is to mark them out on a bit of white paper, and if you are drawing on tracing paper or cloth this is the only way. You can do a complete set at once and so theoretically save a lot of time. I find that one is apt to get muddled and I greatly prefer to use my dividers and do each locus separately. When you run the spline round the level line you may find that your original line as shown in Fig. 6 on page 29 is quite correct, but a few alterations may be necessary. This line governs the shape of the top-side and the appearance of the yacht above water depends largely upon its symmetry. Do not try to distort the curve produced by the spline in order to hide a lack of sweetness in the line. Get it correct even if you have decided that the beam of the yacht is to be 6 ft. and this alteration will add an inch. What does it matter? If you are designing to order and 6 ft. has been stipulated, do not advertise the fact that the design makes it a shade more or less. In any case yacht building is not an exact science like making a car. The ordinary builder will most certainly be a little out here and there even if his moulds are strictly to scale. My *Vindilis* is an inch wider than she should be and she was built by a first-class builder.

Finally complete the design by running in the rest of the level lines. The top one shown in *Sinah* does not reach to the stern owing to the sheer of the hull, and a spline need not be used for it. I have generally found that the level lines and the bow and buttock lines are not a happy family in the upper top-sides and there has to be considerable give and take to get complete harmony with them and the two upper diagonals. This is especially the case with a transom or counter-stern. The natural ending for a yacht of the type of *Sinah* is the canoe-stern. The counter fits better on to a yacht with a harder bilge and fleeter section. It is exceedingly difficult to draw a shapely counter for the deep-bodied high-sided hull.

I now like to indicate the paint line of the boot-top. The majority of yachts, I say this without any hesitation, are ruined in appearance either by not having a paint line at all or by having it lined in incorrectly. If the designer gives the builder no indication of the correct position of this all-important line how can he be expected to get it right? He may be, and probably is, an excellent craftsman, but it is not likely that he is an artist as well. Run this line in with a thick spline. The same spline will be used to put in the gold-line or the rubbing strake, whichever is decided upon. It is usual in this situation to carve a deep groove known as a "cove" to hold the gold. It has always seemed to me a very foolish proceeding, for it weakens the sheer strake very greatly. I introduce my gold-line very early in the designing process, for it looks better if the construction curves do not intrude into the top strake region of the body-plan except as pecked lines.

The design is now complete and we must ink it in. Before this is done I most strongly recommend the beginner to *pencil* it in with an HB pencil and then to put the board up for a few days' study before taking the irrevocable step with Chinese ink. Often a line may appear to be slightly unfair or not sweet, and it may be that only the thickness of a line is necessary to make it look correct. Never forget that a curve is strongly influenced by its surroundings. Perhaps you do not like the look of one of the sections. Trace it separately and you may find that it is perfectly sweet and correct. If you have already gone over the design with an HB pencil you will have no difficulty in inking in. Do not overfill the pen, and keep it scrupulously clean. *Never remove a spline till you are quite sure that the ink is dry*. Be most careful in joining up curves on the body-plan. This is a matter of practice. Later on when you have completed the cabin plans you may add the deck-house, mast, and other deck structures as I have done in the design of *Sinah*. Now have a good clean up. Yacht designing is an unclean job, and unless you are very careful to clean your tools there will be a lot of filth to remove. The scales, curves, set-squares and T-squares seem to attract dirt, and the drawer in which they are kept ought to be constantly cleaned. The dirt is nothing but black-lead from the pencils and one wonders why there is so much of it. Before inking in let me again warn you to remove all fragments of india-rubber from the paper. It becomes electrified in dry weather and sticks to all celluloid instruments, and to some extent even to wood.

The design is now complete as far as its form is concerned. So far it has been drawn solely by eye and common sense. We know nothing about its weight, where that weight is centred, how it is distributed, or about the position and weight of the lead or iron keel, but we do know that our hull has an harmonious appearance, and that the metacentric analysis cannot be far wrong.

CHAPTER IV

The Calculations

WE must now proceed to the calculations, and as I have already said these are quite simple.

The most important is the displacement or total weight.

Whereas a floating body displaces its own weight of water, it is obvious that if we can find out the number of cubic feet contained in that part of the hull which is under water, we have obtained the number of cubic feet of water displaced by the hull. A cubic foot of sea water weighs 64.3 lb. and of fresh water 62.5 lb. In round numbers we say that there are 35 cubic feet in a ton of sea water and 36 of fresh water. I do not happen to have by me the calculations of *Sinah*, so I must introduce you to yet another yacht, *Paida*. The name *Paida* has no meaning. It was intended to be *Paidon*, which is the Greek for a small child, but in some mysterious way during the lettering this became *Paida*, and so *Paida* it must remain. She represents the cruiser of the post-war era when purses will be shallow. She is 19 ft. overall and 16 ft. on the L.W.L. She was intended to be 6 ft. in beam, but she grew nearly an inch in fairing. Before we take leave of *Sinah* I must tell you something about her. I knew very little about the metacentric theory when I designed her, but, as I have said, I tried to follow the lead of yachts designed to this system. I balanced all the sections on pins at the three extremities, and thus obtained their centres. These I referred to the middle-line of the body-plan and obtained the curve, labelled "the curve of centres of heeled sections." It is, as all can see, perfectly symmetrical. Then came the turning-point in my designing life. Admiral Turner asked me to come and spend a week-end with him and I took the design of *Sinah* with me. He showed me how to do an analysis, how to poise sections on a razor edge, and many other useful things. We found that *Sinah* was a perfect example of metacentric balance, and also a metacentroid. I shall explain this term later. Admiral Turner was at that time living in Sinah Lane hard by Sinah Lake, hence the name I have adopted. The design has been built to by Mr. E. F. Hingeley, then living at Copenhagen, and the boat is now in Sweden. She turned out to be an ideal sea-boat, perfectly balanced.

Paida is a two and a half tonner, to be exact 2.49, and she is drawn to a scale of an inch to a foot. Before I go any further I will explain what we mean by Thames tonnage. It is obtained by the following formula:

$$\frac{(\text{L.B.P.}-\text{Beam})\times\text{Beam}\times\frac{1}{2}\text{ Beam}}{94}$$

L.B.P. is length between perpendiculars, that is from the fore side of the stem to the after side of the stern post. In the case of *Paida* we have:

$$\frac{(19-6)}{94}\times(6\times3)=2.49$$

This formula had some meaning in the days when beam was not as great as today and when there were no overhangs. Take for example a quay punt. Her Thames tonnage would be approximately her displacement. On the other hand we have yachts with long overhangs and plenty of beam whose Thames tonnage is very high, and the internal accommodation correspondingly low. Today Thames tonnage means nothing. The American method of describing yachts by their overall length is equally futile. It would not be difficult I think to agree upon a simple formula which would classify yachts in a manner which would make the tonnage or rating have a definite relationship to actual size. But we are wandering from the point.

Let us place the drawing of *Paida* on a board and get out our planimeter. In the first place realize that the ordinary small planimeter is not an accurate instrument, and that we must check its reading whenever we have the opportunity. The cheapest instrument reads to a scale of one inch to a foot only, and this is quite good enough for our purpose. First test it for general accuracy. Draw a square with 2-in. sides. Run the pointer round the edges of this square and the recording drum ought to read 4 sq. in. If the reading is incorrect do it again and yet again and the mean ought to be 4 sq. in. If it is not the appliance should be returned to the maker for adjustment, otherwise a correction will have to be made for every reading. We will assume that it is correct. In using the instrument keep a finger on the arm to make a gentle pressure on the drum. I always think that a planimeter is a wonderful thing, by a combination of sliding and rolling the scale finally records the area. The main drum is divided into ten parts, and each division again into ten. On this drum we read the unit and the tenths. For example 4.5 sq. ft. If the reading is not quite on the 0.5 we look up to the vernier, note which two lines correspond, and read say 4.53. So the vernier reads the

PLATE 3

(left) *La Bonne* (21′ LWL), La Bonne design. Built 1919 by H. Gale, W. Cowes, for H.J. Booker. She is still Solent-based.

(right) *Davinka* (22′ LWL). Built 1936 by C.A. Fox & Son, Ipswich, for Dr Addey. Has cruised in the Mediterranean, French and British waters and is now East Coast-based. Bogle design (which was made in 1933 for the Little Ship Club designing competition). The design was named after daughter-in-law Joan Bogle Butler (neé Hickson). Six Bogles were built in the UK, but *Tradewind* is now in the Caribbean; *Caracole* has cruised as far as Turkey and back.

PLATE 4

(left) *Mat Ali* (29′ LWL). Built 1936 in Selangor, Malaya for E.J.C. (Peter) Edwards, she cruised extensively in Malayan waters before being shipped to Port Said and sailed for England. During the war she was used by the Dutch Navy. *Mat Ali* has cruised in the Mediterranean and North European waters and is now Solent-based. Original Khamseen design.

(below) *Dorothea* (29′ LWL). Built 1934 by Anderson Rigden & Perkins, Whitstable, for L.J.F. Irving to original Khamseen design. When owned by Peter Tangvald, she sailed round the world but subsequently sank after a night encounter with a floating object. PT took to the dinghy.

hundredths of the square inch. The circular graduated disc by the side of the drum reads in tens. Thus we may read a 2 on the disc, a 4.2 on the drum, and a 3 on the vernier. The area is 24.23.

We have now to find the area of the underwater part of each of the sections on the body-plan. As this shows only half the body, go round each section twice with the planimeter. Write all these readings in a column, and then plot them on a suitable scale on to a line 16 in. long, the length of the L.W.L. This has been done in the case of *Paida*, Fig. 12, and labelled "upright areas." It may be that when you have plotted out these ordinates and run the spline round them you will find that one or two of them do not fit the true curve. Return to the body-plan and measure up again on these doubtful figures. You will always find that there has been an error, for if the yacht's hull is fair the curve must also be fair. In the case of *Paida* I found that one-fifth scale produced a practical curve. Each area was multiplied by 2/10 and laid off with the scale of the design, one inch to a foot. Ink in the curve of upright areas and run the planimeter round it. It will be best to divide it up into two parts for this purpose in front of section 5 and aft of it. Add the two together, multiply by 5, and you have the displacement in cubic feet. Divide this figure, which in the case of *Paida* is 64.3 cu. ft., by 35 and the quotient is 1.837 tons, which is the displacement of the yacht. So much for the fearful calculations! One simple division which an intelligent child of seven could do. You will note that the curve has given a useful check upon the inaccuracy of the planimeter.

I have for many years taken the width of the stern post into my calculations for displacement. On mature consideration I think that this is an incorrect practice unless the stern post is almost vertical. On the other hand it may be good practice to include the rudder in the displacement. If we do this, the curve I have drawn is near enough for all practical purposes. Now trace the curve on thick tracing or detail paper, fold it longitudinally, and poise it on a razor. This will give you the position of the centre of buoyancy as far as its fore and aft position is concerned. It is marked on the sheer plan of the yacht with an arrow labelled CB. On the base line of the curve it is marked in as UB, upright centre of buoyancy. Later on I shall introduce you to HB, the heeled centre of buoyancy. The only absolutely necessary calculation that we still have to do, and it can be a very tiresome one, is concerned with the design of the lead keel. I shall introduce you to another graphic method for finding the displacement and simple calculations. This will give us a triple check on the result.

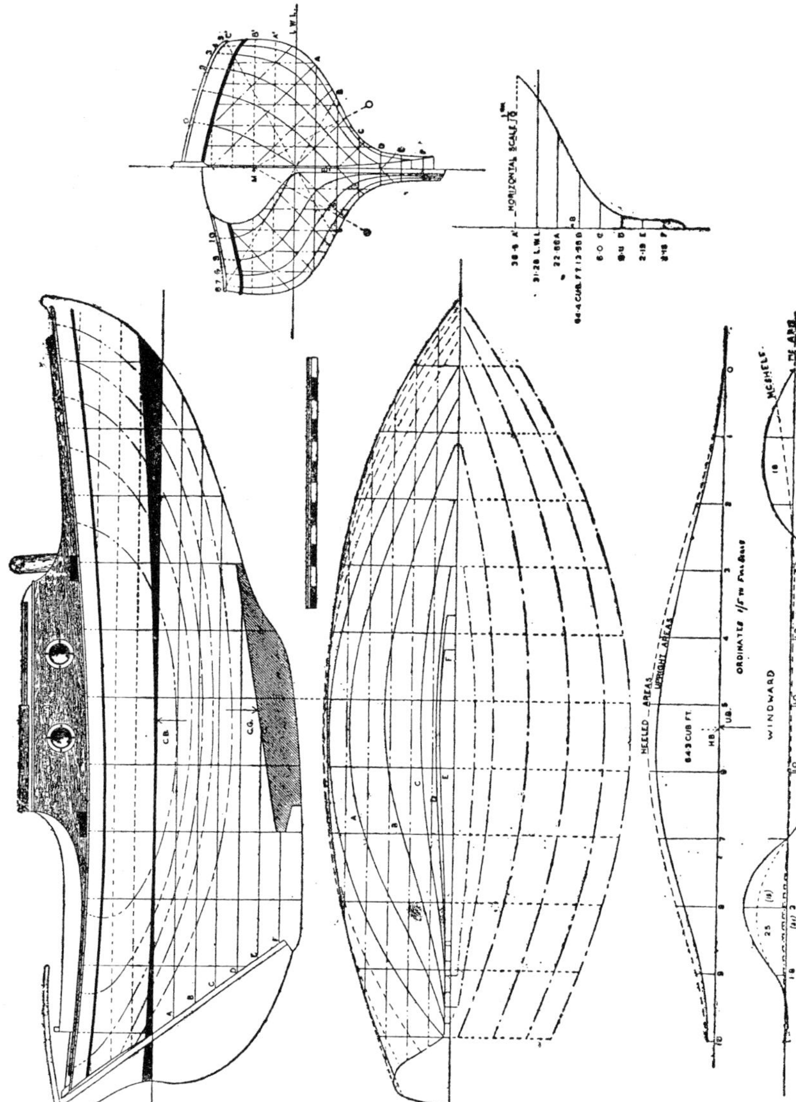

Fig. 12.
PAIDA.

T.M.
$2\frac{1}{2}$ tons.
L.O.A. 19 ft.
L.W.L. 16 ft.
Beam 6 ft. 1 in.
Draught 3 ft. 6 in.
Displacement 1.84 tons.
Lead keel 14 cwt.

In the previous chapter our design was completed as far as the hull is concerned, and we had obtained the displacement by the use of the planimeter and one simple division sum.

We obtained our displacement by measuring up the area of our curve of upright areas with the planimeter, and we found the upright centre of buoyancy by poising this area on a razor.

This will be a good time to discover the heeled centre of buoyancy. We have already ruled in the two inclined water-lines, and utilized them with the aid of the Welch axis to make a preliminary estimate of the metacentric axis. They will now be employed to enable us to plot out a curve of heeled sections. If two body-plans have been drawn, a double bow and a double stern, then it is quite simple to run the pointer of the planimeter round the sections and refer the areas found to the drawing of the curve of upright areas (see Fig. 12) and so superimpose upon it the curve of heeled areas. But in our case we have drawn only one body-plan so we must run our planimeter round in the following manner. Referring to Fig. 13 which shows sections from the fore and after body; begin at A, run up the inclined water-line to B, then go to D and follow on to C and E, returning to A. Then down again to D, on to C and E, and finally come up to the finish at A. For the after section; begin at A, drop down to E, out to F, up to G through H, and drop down to A. Then down to E again, then F, up to H and finish at A. The courses are: ABDCEADCEA and AEFHGAEFHA. You now have the complete areas of two sections. Treat all the others in the same way.

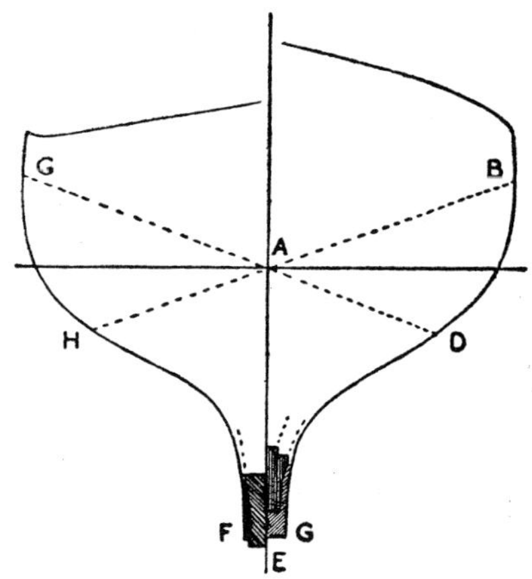

Fig. 13.
Sections from the fore and aft body.

Note that when computing the areas of the upright hull we went round twice because we were dealing with half-sections. Now we go round the complete circuit only once because we are dealing with two portions of the whole heeled section. When this curve has been plotted out with reference to the same axis as that of the curve of upright areas, it will be noted that our new curve is larger than the original. Therefore when the hull has heeled it has become considerably heavier. This is, of course, impossible, so the whole hull will rise till the two curves are equal in area. That is to say that our curve is not really the true curve of heeled areas. However, the force of wind that heels the hull will to some extent counteract its tendency to rise, and therefore the actual difference will not be as much as is shown by the curve.

In practice it is convenient to assume that the hull does not rise. The actual amount of rise will depend upon the character of the midship section. The Norwegian type with flaring top-sides will develop a large lift of hull on heeling, and it is this feature which makes them uncomfortable and seasick-making. A steamer, on the other hand, with underwater low-placed bilges and with considerable tumble-home, may not rise at all. A comparison of the two curves in *Sinah* and *Paida* shows a considerable difference. *Sinah's* curves tend to approximate more than those of *Paida*, the lower softer bilge of *Sinah* being responsible for the difference. The next step is to cut out a template of the new curve and poise it to discover its centre. This is the H.B., the heeled centre of buoyancy. In *Sinah* they practically coincide, whereas in *Paida* they are very close together, the separation being 0.047, a negligible figure.

If the heeled centre of buoyancy lies aft of the upright centre, it is obvious that the displacement of the fore-body has decreased at the expense of the after-body, the two are no longer in equilibrium and so the hull will root by the bow till the balance has been rectified. The vast majority of the older type of cruising yachts and fishing craft round our coasts have this fault, and it is one of the causes of griping. The old Itchen Ferry type of fishing boat and yacht was an extreme example of a hull with wide separation of these centres. Like Mr. John A. Stewart of Clynder, I was once greatly struck by the lines of one of these vessels, *Foam II*, when I saw them in Robert C. Leslie's book, *A Waterbiography*. My friend, the late L. Boughton Chatwin, was so impressed with her that he built a model. Alas! when put on a wind in a strong breeze she heeled to her deck line, and rooted to such an extent that she put her bows under and refused to go to windward at all. To bring these centres together must be one of the aims of a designer, because in so doing he has gone a long way towards obtaining good balance. The matter is fully discussed in Chapter X. It is obvious that if the centre moves aft on heeling, this defect must be remedied by filling out the shoulders and fining in the quarters.

Although I have shown how the displacement can be obtained without calculations, I think it will be well to indicate how these are carried out. There are two methods, the Trapezoidal rule and Simpson's rule.

To find the area of a curve by the Trapezoidal rule we draw a line across it, in the case of our own calculations the base line of the curve of areas, divide it up into equally spaced sections, in our case ten, and at each point raise a normal or ordinate. The sum of the lengths of these ordinates, less the sum of half the terminal ordinates, multiplied by the spacing will give the area. If we use millimetres for the measurements, the area will be expressed in square millimetres; if we use inches the area will be given in square inches, and so on. In the case of *Paida* the upright areas in square inches obtained by the planimeter are arranged in the following order:

0.00+0.80+2.50+4.34+6.00+7.20+7.36+6.05+3.95+1.66+0.76=40.62 (sum of areas or ordinates). Deduct 00.38 (half of the last area)=40.24 (remainder when the half area has been subtracted)×1.6 (the spacing of the sections in inches)=64.38 (cubic feet).

The Trapezoidal rule gives a result that is slightly less than the correct figure, for it assumes that the ends of each ordinate are joined by a straight line and not by a curve. In the case of a curve of displacement which ends at each extremity in a reverse or concave curve, the area obtained would be about correct.

The second method, by Simpson's rule, was, I believe, invented by a weaver about three hundred years ago. To use this rule it is necessary that the area to be measured shall be divided into an even number of parts, and that is a very good reason for dividing the load water-line of a design into an even number of divisions. We arrange the areas of the sections or the lengths of the ordinates in a column, and proceed to multiply them by Simpson's multipliers. The first is multiplied by 1, the second by 4, the third by 2, and so on till we come to the last which is multiplied by 1. The products are added together and the sum divided by 3. The quotient is multiplied by the spacing of the sections, in the case of *Paida*, 1.6 ft., and we obtain the number of cubic feet of the displaced water. Divide as before by 35 and the quotient is the displacement in tons. Do not ask me why we divide by 3, for I have not the vaguest idea. It is, I believe, part of a mysterious process called integration.

Here we have the calculations in tabular form:

	Area of sections		*Simpson's multipliers*		*Products*
0	0.00	×	1	=	0.00
1	0.80	×	4	=	3.20
2	2.50	×	2	=	5.00
3	4.34	×	4	=	17.36
4	6.00	×	2	=	12.00
5	7.20	×	4	=	28.80
				Carried forward	66.36

	Area of sections		*Simpson's multipliers*		*Products*
			Brought forward		66.36
6	7.36	×	2	=	14.72
7	6.05	×	4	=	24.20
8	3.96	×	2	=	7.92
9	1.66	×	4	=	6.64
10	0.76	×	1	=	0.76
					120.60

$$\frac{120.60 \times 1.6}{3} = 64.32 \text{ cubic feet. } \frac{64.32}{35} = 1.837 \text{ tons}$$

This displacement is the same as that obtained by the far simpler method of measuring the area of the curve of upright areas with the planimeter.

We obtain the centre of gravity of the curve of upright areas by tracing it upon a piece of detail paper or tracing paper, cutting it out with scissors and poising it on a razor. In actual practice before doing this we first draw the curve of heeled areas, trace it on the same paper and cut it and poise it. This gives the centre of the heeled areas. Then we trim it down with scissors to the curve of upright areas and poise this as before. It is not necessary to make two tracings. There are some, especially among the Olympians, who profess to scoff at the poising method, so perhaps I had better show you how to calculate the centres of these and other curves.

Until quite recently I used to consider the areas of the sections as weights hung along a lever and calculate their centre by the usual method of taking moments. This was the plan described by a most eminent designer in some articles written about designing, and I used it without any idea that it was fallacious. I never could discover why when I was working out the centre of lateral resistance I could not harmonize two methods. Then when I took to poising I found that there was a constant error which placed the centre of gravity of a metal keel about 3 per cent. further aft than it ought to be. The fact is that the areas of sections of say a lead keel cannot be treated as though they are weights, for this takes no account of the material between the sections. In the case of a displacement curve coming to a point at each end, this method is sufficiently accurate, but with a wedge-shaped article like a keel it is definitely inaccurate. I regret that the centres of the keels marked on the majority of my designs, and I suppose equally so in the work of the eminent authority mentioned, are marked too far aft. It is fortunate that the error is in this direction, for it is dreadful to get the weight of a ballast keel too far forward. The centre can be obtained by using the Simpson rule or by a modification of the Trapezoidal rule.

Going back to the last table of calculation by Simpson's rule, take the products given when the areas have been operated upon by Simpson's multipliers. Set them out in a column and multiply them by 0, by 1, by 2, by 3 and so on successively. Add both columns, multiply the last by the spacing of the areas, and then divide the product by the sum of the first column. The quotient will give the distance of the centre of buoyancy from the fore end of the L.W.L. as below:

0	0.00	×	1	=	0.00	×	0	=	0.00
1	0.80	×	4	=	3.20	×	1	=	3.20
2	2.50	×	2	=	5.00	×	2	=	10.00
3	4.34	×	4	=	17.36	×	3	=	52.08
4	6.00	×	2	=	12.00	×	4	=	48.00
5	7.20	×	4	=	28.80	×	5	=	144.00
6	7.36	×	2	=	14.72	×	6	=	88.32
7	6.05	×	4	=	24.20	×	7	=	169.40
8	3.96	×	2	=	7.92	×	8	=	63.36
9	1.66	×	4	=	6.64	×	9	=	59.76
10	0.76	×	1	=	0.76	×	10	=	7.60
					120.60				645.72

$$\frac{645.72 \times 1.6}{120.60} = 8.56 \text{ feet}$$

Naturally this set of figures will in practice be combined with the last set, but I have separated the two in order that the calculation for displacement may be displayed in all its native simplicity. If you now carry out the same computations for the heeled areas you will arrive at the centre of buoyancy of the heeled areas. But the whole thing seems very futile when poising is so accurate and simple. I have, however, given these figures because they will be of value in the computations for the lead keel.

CHAPTER V

Ballasting

WE have completed all the calculations absolutely necessary for the hull itself, and have now to tackle the ballast keel. This is a subject that is glossed over in most articles on yacht designing, and the books on naval architecture do not give the matter the attention it deserves.

The ballast of a small sailing yacht will in the vast majority of cases wholly or in part be carried outside the hull in what the Americans call a shoe. I have made it quite plain that the smaller the yacht the greater the necessity to obtain as much stability as we can. One factor in stability is outside ballast carried as low down as possible. In the case of the "tabloid" cruiser it is essential to get it almost all outside. In a larger ship, say a 7-tonner, we can place some inside, and in a still larger craft it is advantageous to place even as much as a third inside. There is, however, a bogey to lay: it is held, and with some reason, that outside ballast makes an uncomfortable yacht, but this does not apply to fore and aft stability.

A yacht with outside ballast will not pitch any more than one with it all inside, but the period of roll is definitely affected by the position of the centre of gravity of the hull. I once made a voyage to the Cape in a Castle liner loaded with locomotives and other heavy stuff low down in the hold. The effect in a December gale, the gale in which Old Harry lost his wife, was devastating. Off the Burlings we developed a roll of 64 degrees, and one of a most unpleasant character. The captain of a Union Castle liner in which I was doctor used to have the ballast tanks pumped out to ease the roll. This is a matter of importance in steamers, and in large sailing yachts, where too much initial stability may cause a jerky motion that may put a great strain on the spars and the crew. In a small yacht we are not concerned with comfort. The sails keep the craft from rolling and in heavy going things are so uncomfortable that a little more or less would not be noticed. In any case we have got to have the stability. Longitudinal stability is influenced by the fore and aft distribution of the ballast. If it is too much spread out the yacht may pitch heavily. If too much concentrated she will be too lively and will invite sea-sickness.

One of my designs was built of steel in Holland. The keel was in the form of a trough in which the lead ballast was stowed. The owner took his father out for a sail and he got very sea-sick and refused to go again. The ballast was spread our more fore and aft, and then the yacht became a diver and took it green over her bows, but the happy mean was discovered; the yacht became dry and buoyant and father again became a passenger. Probably the ballast in a small sailing yacht should occupy about one-third of the length of the L.W.L. If it be of iron it will have to be carried out further at each end to obtain the necessary weight. Are we to use lead or iron? If expense is of no object have lead every time.

It docs not by any means follow that the market price of lead and iron is the sole factor in the total expense. The use of iron may give rise to more expense in the actual hull construction. For example in the Z 4-tonners an iron keel was specified because at the time she was designed lead was fetching an abnormally high price. Her underwater profile has a good rise for'ard with the effect that in the sections for'ard of amidships the keel becomes very wide if iron is used. This wide keel has to be cut out of a large bit of elm, and whereas it tails off to a very narrow width at the stern, there is much wood cut to waste. If the keel were of lead this wood keel would be lower and narrower for'ard, and the wood necessary would cost considerably less. Again the average builder casts his own lead keel close to the yacht that is being built. The iron keel has to be cast at a foundry. This means that the mould has to be taken to the foundry, the casting has to be brought back, and then the profit of the foundry has to be met. So in many cases the lead keel will not cost much more than iron if the market price is not abnormally high.

For small yachts lead is best. It can be fitted more closely because the top can be planed to a good fit. The bolt holes can be bored *in situ*, and any necessary adjustments made. On the other hand if lead is used and brass bolts are employed, then these bolts must not be fitted through steel floors. If a brass or bronze bolt is brought in contact with a galvanized steel floor it is only a matter of months before galvanic action eats away the zinc and then attacks the iron. Far rather would I use iron bolts for a lead keel than metal bolts through the floors. The difference of electric potential between lead and iron is small and there is not much galvanic action. On the other hand the difference in potential between brass and zinc is high and electrolysis is active. Have a lead keel, naval brass bolts, and floors of forging bronze (Dixtrudo). This costs a little more but is lasting. The other homogeneous combination is the all-steel one. Iron keel, steel or better iron bolts, and steel floors. Do not wreck this homogeneity by using galvanized bolts! The bolts should be heated and quenched in linseed oil. The final objection to iron is that it cannot be altered. In the case of a small yacht if the lead is too heavy or if the trim is wrong, it is easy to saw off a little from one end or the other. In some cases two spaces are left in the top of the lead keel fore and aft and filled with wood. If the yacht is above her L.W.L. these blocks of wood can be replaced with lead either at both ends or at one or the other to correct trim.

What proportion of the total weight of the hull should be in the ballast? My rule in the case of a small cruiser that has to be filled up with all kinds of weighty gear and cruising duffle, was to multiply the total displacement by 0.4. If the construction has been carefully thought out and kept as low in weight as is consistent with adequate strength, then this will be about correct, but in the majority of cases it is too much. I would make the factor 0.35 in the small cruiser, say up to 6 tons or so, and in the "tabloid" I would keep to 0.4 and cut some lead off if necessary. Over 6 tons we can put some of the ballast inside, and we are out of our difficulties. The day boat can have considerably more ballast outside.

Not only have we to make our keel of the correct weight, but we must so fashion it that its centre of gravity is correctly placed. There is a proper way to obtain this position and also the correct weight of the keel. It is difficult, and tedious, and one professional designer told me that it would not pay him to do it in the case of a small yacht. You work out the weight of everything in the yacht; skin, beams, deck planking, engine, anchors, etc. Thus you obtain the weight of the hull, and knowing the displacement, the difference is the weight of the ballast. This is bad enough but there is worse to follow. You now take moments of all these weights round any given point on the L.W.L. and so find the position of the centre of gravity of the hull in the fore and aft direction. Let us now complete the horror; if you want to obtain a curve of stability you begin again with the weights, and take moments in a vertical direction above and below the C.B. This will give you the centre of gravity in a vertical sense, so many inches above the C.B. or more rarely so many below. Of course, you have to assume that the wood employed has its scheduled weight and that this weight is constant for all the wood used.

One racing yacht built a few years ago was down by the stern and drove her designer to distraction till it was found that the oak used in the stern frame was 50 per cent. heavier than that used in the fore part. Another 20-ton yacht was launched many years ago and she floated some inches above her designed L.W.L. Calculations were checked over and over again, always with the same result. Years later a yard hand confessed that the lead mould had been filled with steel punchings and lead poured round them. In the best and most careful hands these calculations lead to inaccurate results, and it is a common thing for the yacht when launched to be above or below her bearings. Three yachts were built by three different firms to one of my recent designs. One was down by the head, another by the stern and the third floated level, but needed quite a lot of inside ballast to put her down to her marks. So you have two disturbing elements: the calculations can be and often are wrong, and on the other side the builder does not always build exactly to the drawings or use the same materials.

It comes to this: it is practically impossible to guarantee that a yacht when launched will float to her designed L.W.L. In the case of the very small craft it is wise to err on the side of being too heavy, for it is easy to trim the shoe. In larger craft err on the other side for a little inside ballast in immaterial.

We must arrive at some simple compromise. I am sure not many of us are going to tackle these calculations, in fact few of us could find the time or have the necessary experience and knowledge.

In Chapter XII Commander Braithwaite has explained all the details of the method for weighing the hull and locating its C. of G. The full calculations will be found in Skene's *Elements of Yacht Design*, page 68. The method I adopt is as follows: I multiply the displacement by 0.35 and adopt the product as the weight of the ballast. Then I assume that the weight of the hull about corresponds with the centre of buoyancy. I then consider the weight of the engine and balance that against the weight of the chain. Thus in *Vindilis* the engine weighs about 200 lb., and the 30 fathoms of ⅜in. chain 240 lb. If the chain locker and the engine are at approximately equal distances from the centre of buoyancy the two will balance each other. The weight of two men aft in the cockpit, say 400 lb., can be balanced against the weight of the mast and the anchor alongside the mast. The centres of any two weights can be ascertained either by taking moments, or by the easier method that we adopt for finding the centre of effort of the sails. I shall describe this later on.

These are very crude considerations and it boils down to the position of the lead keel being placed about 1.5 per cent. of the length of the L.W.L. in front of the centre of buoyancy. The error must be in the direction of having the weight too far forward for it is far easier to stow trimming ballast aft than for'ard, and it is also stowed lower down.

I think that in a yacht of about 6 tons, if we multiply the displacement by 0.35 and fix the centre of gravity of the keel about 1.5 per cent. in front of the centre of buoyancy of the hull, we shall not be far wrong, and we shall have allowed for a little trimming ballast. Any tanks aft must be taken into consideration, but care must be taken to keep weights of this sort as far as possible amidships.

We now come to the actual method of designing the keel. There is only one, the very old-fashioned one of trial and error. You may hit it off first time, but it may take many hours. Of course, the more experience one has the nearer the first guess will be to the final result. Take the sheer-plan and rule a line which you imagine will be the top of your ballast keel line. Rule a vertical line aft to mark the end of the keel. The forward end will generally be where the top of the keel meets the lower edge of the stem. The keel will thus be wedge-shaped. You will now look at this keel and consider that the sections for'ard are smaller than those aft and that, therefore, the centre of gravity is aft of the centre of the keel. Make as good a guess as you can and then rule in the tentative keel. Now turn back to Fig. 10 on page 42. Neglecting the butt at the after end there are five sections implicated in the keel: 5, 6, 7, 8 and 9. The first section at 5 is a very small one. In the original drawing it was larger, but it has diminished in tracing and reproduction. Take your dividers and measure downwards from the water-line on the body-plan the distance from the top of each section of the keel. This has been done in Fig. 13, and each keel section so obtained has been shaded in a distinctive manner. You will note that there are two black ones, one vertically striped and the other two ruled diagonally right and left. With the planimeter obtain the area of each of these sections. The planimeter is not very accurate with such small areas as that of section 5, but if you take a mean of three or four readings you will be near enough. Now treat these readings as we did those we took of the vertical sections to calculate the displacement. Either compute them or make a curve and take its area with the planimeter.

Before going any further I will suggest that if your planimeter reads only to the scale of an inch to a foot you treat the keel as though the whole yacht were drawn to the scale of an inch to a foot. In this case her L.W.L. will be 18 ft. Take these five figures, add them up, deduct half the sum of the two terminal figures, and multiply by 18. This will give you the number of cubic feet in the tentative keel. Multiply this by $4^3/3^3$, that is 64/27, because *Sinah*'s scale is $\frac{3}{4}$ in. to 1 ft., and you have the number of cubic feet in the keel. Multiply this by 710 because a cubic foot of cast lead weighs 710 lb. (a cubic foot of cast iron weighs 450 lb.), and you have the weight in pounds. Now draw a curve of these five figures, cut it out and poise it for the centre of gravity.

If you are very lucky the result will be that the weight is about 2 tons, and the centre of gravity 7 ft. aft of section 5. It is very unlikely that you will have this luck. Perhaps the weight is right, but the C. of G. is too far aft or for'ard. If the error is slight it can be corrected by shifting the whole keel fore or aft, but if material you will have to alter the slope of the top till the answer is what you require. Say that the C. of G. is right but the weight wrong. Then raise the top or lower it till this is corrected. There are two variables, and time, patience, and again patience are called for. Finally the result is correct, but we have not finished our task. The builder will want a drawing of the keel, and the work we have done on our original drawing cannot be accurate. The scale is too small and the sections too few.

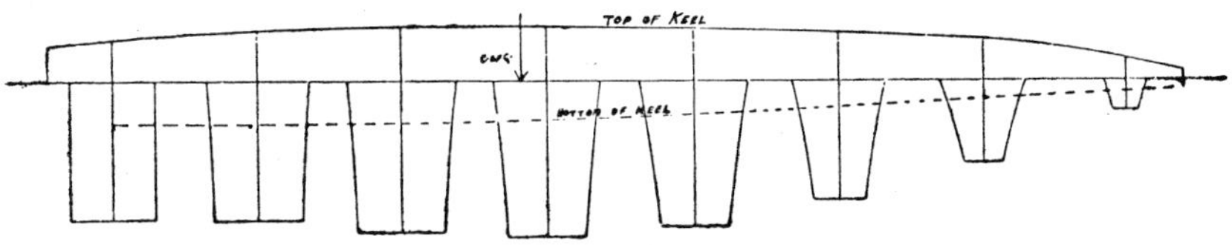

Fig. 14. The drawing of *SINAH*'s keel.

Fig. 14 is the original drawing, or rather tracing, from which the blue-print was made that was sent to the builder of the first and, as far as I know, the only *Sinah*. There is a tale attached to this keel which has an important implication. When the keel was cast in Denmark it weighed not 2 tons as calculated, but 2.4 tons. This was a large error and my name would have been "mud" had not the yacht, when launched, floated rather above than below her designed L.W.L. Here was then not so much an error as a mystery. I went over the computations for keel and displacement several times, always with the same, the original result. Two years later, when fortunately the yacht was sailed from Denmark to Sweden, the owner in a talk with the builder discovered that no allowance had been made for the thickness of the planking in making the moulds. This has happened to two other yachts that were built to my design. In olden days it was the custom to make the design not to the outside of the planking, but to the inside, and builders took the sections shown on the body-plan as those of the moulds or frames. Of course the finished *Sinah* with an added beam of 2 in. had a very much larger displacement than she was designed for, and her keel was also heavier. The lesson to be learnt is that the builder should in all cases be furnished with a table of offsets. In Chapter XIII I shall describe how this is made. The bottom of the keel is made by measuring the widths of each section at the heel, doubling them, and plotting them as a curve. The same is done for the top of the keel. The sections are spaced at double the distance they occupy on the main design. The scale of *Sinah*'s keel is therefore not $\frac{3}{4}$ in. to a foot, but $1\frac{1}{2}$ in. to the foot. The keel is double the length of the actual keel and double the width and depth. It weighs, of course, eight times the weight it would do at the correct scale. Keep this simple fact in mind, for when we come to find the weight of this double-size keel we shall have to divide this weight or cubic capacity by eight. When you come to draw the top and bottom of the keel you may find that the curve is not a sweet and harmonious sweep. In this case small alterations will have to be made to the heels of the sections till all is ship-shape. Now interpolate an additional station between the primary sections, and from the measurements across the bottom and top of the keel draw a new section between each of the primary five sections.

You now have eight sections and seven spaces between them. This means that we cannot employ Simpson's rule for the calculations. These are not absolutely necessary, but desirable; I showed you that the *computation* for finding the displacement was merely to divide 64.3 by 35. We were then weighing water and a small error was of no importance, but now we are weighing lead or iron, and we must employ all the checks we can, both for weight and centre of gravity.

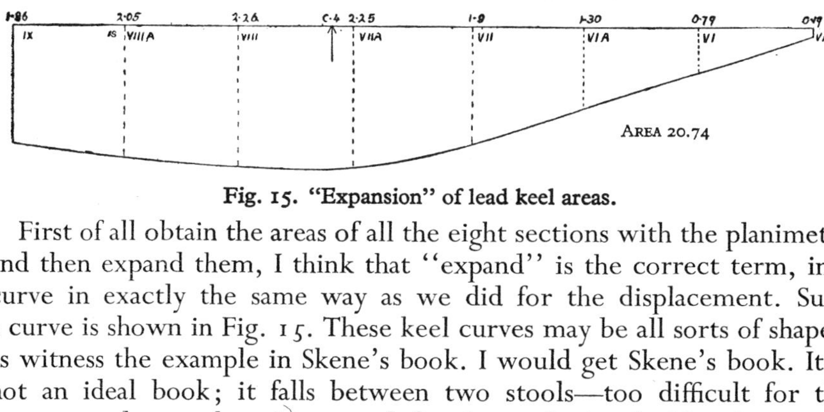

Fig. 15. "Expansion" of lead keel areas.

First of all obtain the areas of all the eight sections with the planimeter and then expand them, I think that "expand" is the correct term, in a curve in exactly the same way as we did for the displacement. Such a curve is shown in Fig. 15. These keel curves may be all sorts of shapes, as witness the example in Skene's book. I would get Skene's book. It is not an ideal book; it falls between two stools—too difficult for the amateur and not exhaustive enough for the professional. Also there is no sort of suggestion that any effort should be made to produce a balanced hull, nor is any method of so doing suggested. All the same it is full of valuable information and is necessary. A good and simple book is A. A. Symond's *Yacht Design* (Edward Arnold).

Now poise the curve and obtain the centre of gravity of the keel, and take its area with the planimeter. The ordinates were measured off with an inch scale, so the reading is the cubic capacity of a keel drawn to double scale for a yacht 18 ft. on the L.W.L. The capacity obtained by this method is 20.74 cu. ft. This is the actual capacity of the double scale keel. Divide this by 8 and we get 2.6 cu. ft., which is the capacity of the keel of the small-scale yacht. Multiply this by 710 and we have the weight of the keel in pounds, and expressed in tons it is 0.824. Multiply this by 64 and divide by 27, and we find that the keel of *Sinah* is 1.95 tons. I am keeping to the same scale and making a final change over because it is a method which is simple and not liable to go wrong. Of course, the correction can be made earlier, but I have not a mathematical mind and easily get muddled by quite simple things of this sort.

And now for the "computations." They are very simple. As I mentioned before, avoid the error of taking the sections as weights on a beam and taking moments in the well-known way. The keel is not a series of weights, and an error of about 3 per cent. will appear if we regard them as such. We cannot use the Simpson method, so we fall back upon the Trapezoidal. Add the figures, taking only half the terminals. Multiply the sum by the spacing, which is 1.8, and you have the cubic capacity of the double-size keel which ought to be the same as that obtained by the planimeter from the expanded keel curve. Now to find the centre of gravity use a formula which was given to me by Admiral Turner. Here it is: Let H be the measure of the *interval* between the sections, and *n* the number of *spaces*. The areas of the sections are indicated by a, b, c, d . . . z.

$$\text{C. of G. from a} = \frac{H\left(\frac{a}{6}+b+2c+3d+4e\ldots+\left(\frac{3n-1}{6}\right)z\right)}{\frac{a}{2}+b+c+d+e\ldots\frac{z}{2}}$$

Computations for weight of lead keel and for its centre of gravity:

Areas					
0.19	×	1/6	=	0.03	$\frac{47.95 \times 1.8}{11.585}$
0.80	×	1	=	0.80	=7 feet
1.30	×	2	=	2.60	which is exactly as
1.90	×	3	=	5.70	given by poising
2.25	×	4	=	9.00	
2.26	×	5	=	11.30	
2.05	×	6	=	12.30	
1.86	×	$\frac{21-1}{6}$	=	6.20	

	0.19	
	1.86	
12.61	2)2.05	47.95
1.025	1.025 half sums of terminal figures	
11.585		

$$\frac{11.585 \times 1.8}{8} = 2.6 \text{ cubic feet}$$

This is exactly the same as the figure obtained by the far easier method of the planimeter described above.

We have now completed the design including the ballast keel. If we are weary in well doing we can now send the design to the builders, and give the order to build, but I think that most of my readers will wish to make the few more simple calculations that will give us the metacentre from which we can form an idea of the stability, and complete the metacentric analysis. In addition we have to design the sail-plan, and consider the lay-out and a few items regarding construction.

CHAPTER VI

Sails

LOGICALLY I ought to continue this book by describing how we obtain the position of the metacentre; how we make the vertical displacement curve, and from it deduce the vertical position of the centre of buoyancy. I should conclude by a description of the complete metacentric analysis. But I am sure that by this time my readers are tired of figures and theory and prefer something more practical. So I shall continue with sail-plans.

The action of a sail with a following or free wind can be understood by all, but many do not exactly understand how a ship can beat to windward, sailing as some put it against the wind. The fact is that when going to windward only a small fraction of the wind force is utilized to drive the ship ahead. Much runs to waste along the surface of the sail; some is wasted by trying to push the hull sideways, and a third fraction is expended in heeling the ship against the downward pull of her ballast and her initial stability.

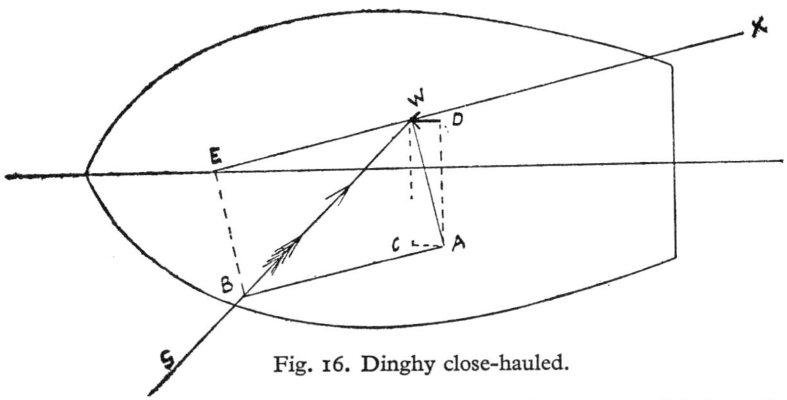

Fig. 16. Dinghy close-hauled.

To explain what happens we must introduce our old friend the parallelogram of forces; Fig. 16 represents a dinghy close-hauled on the port tack lying four points off the wind. I will not complicate matters by introducing the apparent wind, although it is an important subject. EX is the sail and SW the usual SW. wind. We assume that part of the wind stream hits the sail at W. Now we know or should know that any force can be resolved into two forces at right angles to each other. Let BW represent the actual force of the wind. From W draw WA at right angles to the plane of the sail. Join BE and draw BA parallel to EW. Here we have a parallelogram of forces. We have resolved the wind force BW into two forces at right angles to each other, EW and AW. EW does not concern us. It represents force running to waste along the plane of the sail. We are left with AW, a force acting at right-angles to the sail. Note that AW is shorter than BW. We have already lost more than half our wind force. By introducing sines and other complications we can compute how much we have lost and what remains. Now take the useful force AW and again resolve it into another pair of forces at right-angles to each other, CW and DW. CW is useless to us, it merely tries to drive the hull sideways. The small remnant DW is all that remains as an actual forward-acting force. This diagram shows in graphic form how difficult it is to sail to windward and why our ship heels so much more than it does with a free wind. CW is employed in heeling the yacht and trying to cause leeway. Only because the form of the hull is such that it resists leeway and is easily driven ahead can the greatly reduced wind force represented by DW drive the yacht ahead. From this it is evident that the more resistance to leeway that we can supply to the hull the better she will go to windward. Hence deep keels, lee-boards and centre-boards.

There is one redeeming factor that helps a yacht to windward. Running before the wind we reduce its power. If we run before a 10-knot breeze at 5 miles an hour, the actual wind velocity is only 5 miles an hour. On the other hand if we could sail in the wind's eye at 10 miles an hour against a 20-knot wind the available velocity would be 30 miles an hour. This is why an ice-yacht can sail considerably faster than the wind with reaching wind. She is always close-hauled except when running dead.

We must now consider how the wind acts upon a sail. We must think of it as an aero-foil, like the wing of a plane. Careful investigations have shown that 75 per cent. of the force driving a yacht is due to negative pressure behind the sail, let us call it a vacuum for convenience. Only 25 per cent. of the driving force is due to the actual wind pressing on the windward side of the sail. The value of this vacuum is diminished by anything that causes an eddy. A mast is a great offender. The vacuum can be increased by the funnel effect caused by a sail in front of it with a suitable overlap. This is why it is so essential to find the correct lead for headsail sheets. The effect of the eddy-making mast is so great that area for area a sail set behind a stay is twice as efficient as one behind a mast. That is to say that a staysail of say 100 sq. ft. in area is as effective as a mainsail of 200 sq. ft. This is why stowing a staysail give so much relief to a yacht pressed by a hard squall. The fishermen call a staysail a pressing sail, but what they really mean is an efficient sail. Lower 100 sq. ft. of staysail and you have the same effect as taking a reef of 200 sq. ft. in the mainsail. Not only this but you have removed the funnel effect and reduced the efficiency of the mainsail. No wonder the yacht is relieved. This factor also accounts for the modern headsail rigs. More and more of the area is disposed before the mast, and the mainsail diminishes *pari passu*. These considerations make it obvious that a sloop is more effective than a single sail rig. The mainsail standing alone does not get the necessary funnel effect to maintain a good leeward vacuum.

There is another fundamental fact about sail aerodynamics: it is the luff that does the work going to windward and reaching; only when the wind passes to the quarter and gets aft does the after part of the sail possess any value. On other points of sailing it acts by giving a good clearance to the wind from the working luff. You can cut large areas from the after part of the sail without diminishing its efficiency to windward. Douglas Johnson sawed 18 in. off the boom of his *Faraway*. When she was going to windward alongside *Englyn*, her sister ship, there was no difference in speed. Had an equal area been added to the luff of the sail, that is to say, had the mast been lengthened by 3 ft., the speed would have been increased. These facts explain why modern designers try to get as long a luff as they can. In models the most recent designs have immensely long luffs and short booms, but in the real yacht there is a very definite limit imposed by the ultimate strength of material and other factors. For one thing the modern model needs a kicking strap to keep the boom down. Many have said to me: "I have no desire for a fast craft." in the first place this is probably not true, for no one likes all and sundry to give his yacht the go-by, while secondly it may be a matter of life and death to have a yacht that can beat out of a tight corner, and it may be that only a racing type of craft will do it. This is one reason why the Bermudan rig is the best for small craft. When *Sandook*, a gaff cutter I owned at one time, had two reefs down, her luff was very short and that was a most ineffective rig for getting to windward, in fact probably the jib did most of the work. If you carry a gaff rig you must imitate the fishermen, reef the mainsail, but still carry the topsail and so preserve the luff. The old Bembridge rig must have been an effective one for this reason, it was a combined mainsail and topsail, and however much reefed it always had a long luff. The Dutchman's mainsail has a long luff and a short gaff, a most efficient rig, and one devoid of much rigging.

I have, however, already strayed far from the proper subject and will not err further by entering into a discussion on rigs. Suffice it to say that I think the one-sail rig must be a lug, the dipping-lug, though unsuitable for yachts, being most effective because it has an eddy-free luff. The bob-tailed cutter, and short-boomed sloop have made it unnecessary to adopt a yawl rig. The sloop is excellent up to say 5 tons. After that cutters of all sorts, and then the Bermudan ketch. The gaff-rigged ketch is an appalling thing to put to windward, but the Bermudan is excellent, for it has two long luffs. I think that it may for short-handed purposes be adopted for yachts of 10 to 15 tons. The larger yachts, over 20 tons, should be gaff rigged with long topmasts and efficient jib-headed topsails. If you choose a yawl the mizzen must be well inboard and have a sound stepping so that the sail can be carried in all weathers, otherwise it is a delusion.

We now come to the actual design of the sail-plan, and there is practically nothing to say about it, for you may have it as you like provided that you keep in mind what I have said above. If you are designing for one of those yachtsmen who always turn on the engine to go to windward, then you have still more latitude, for any old bag will do with a fair wind. We must, however, work out the sail area, and consider how much the yacht will carry. Some may say that we ought to know something about the centres of lateral resistance and of effort. We will come to this later on. We have used *Paida* and *Sinah* for our designing, and we will now study their sail-plans. Fig. 17 is the sail-plan of *Paida* drawn to a scale of half an inch to the foot, half the scale of the yacht. This is generally convenient. If your design is to a scale of $\frac{3}{4}$ in. to 1 ft. use a scale of $\frac{3}{8}$ in. to 1 ft. for the sail-plan, and so on. *Paida* is a small thing that can easily be carried on a steamer or lorry, and is fitted with a lowering mast for entering all sorts of odd places. For example she can go up the Hamble above Bursledon bridge and explore the beautiful upper reaches of this river, at high tide of course. A long Bermudan mast would be unhandy for these purposes, so I have adopted the next most efficient rig, the one time small racer's rig, the gunter-lug.

This has many advantages and is simple and cheap. The total sail area is rather on the small side, 177 sq. ft. An X-boat with the same water-line and almost the same beam has 200. Well, *Paida* is a cruising X-boat. She has the same weight of ballast as an X-boat, but her hull with cruising gear will weigh more and so she cannot carry as much sail. The yard is fitted with jaws like a gaff, but whereas the yard is peaked high, the jaws must be strongly bent, or better, as shown in the drawing, a hinged-gaff is

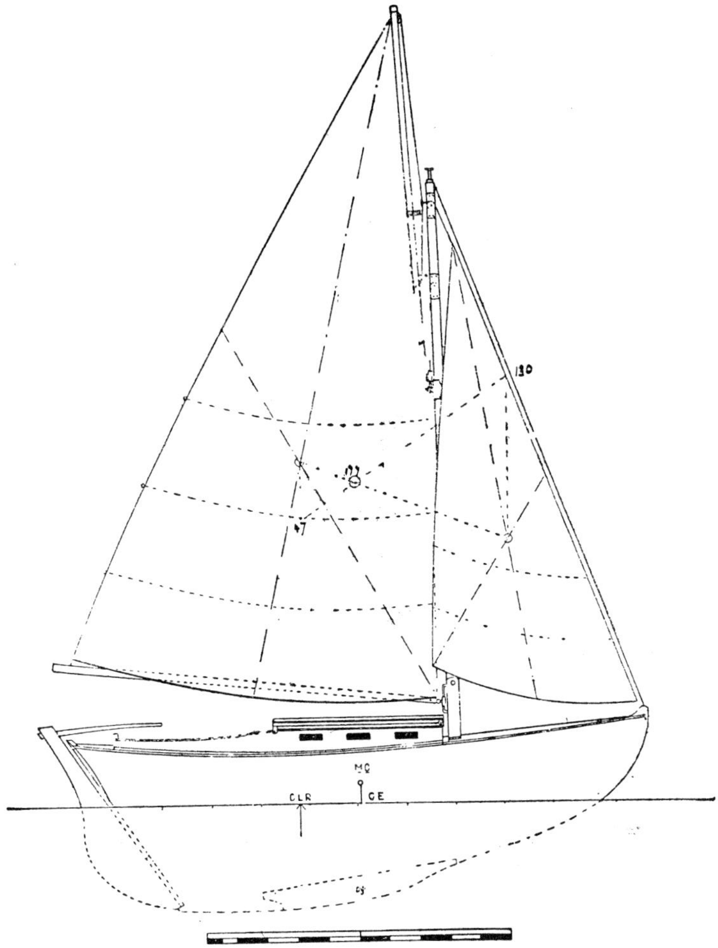

Fig. 17. *PAIDA:* sail plan.

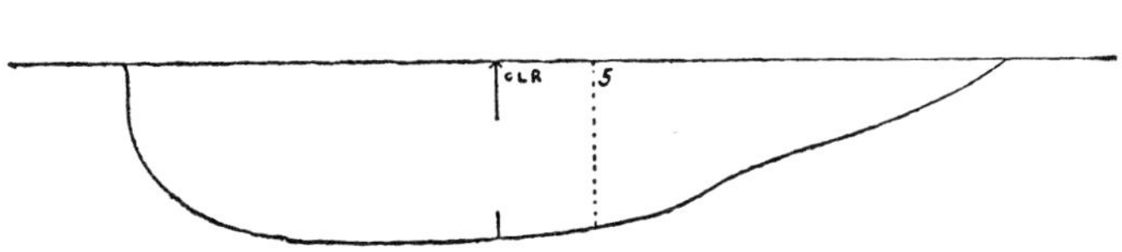

Fig. 18. Under-water profile.

employed. The head of the yard is held to the mast by a wire span. The span must be strong, for the strain on it is considerable and the closer the yard can be got to the mast the greater the tension induced. A specially designed saddle-shaped shackle is bent to the peak halliard, which passes over a sheave in the mast like the topsail halliard of a large gaff-rigged yacht. This is of course not the best mechanical arrangement for the purpose, but it works and is simple and cheap. If the splice shows signs of wear it is easy to turn in a new one. This method gives the minimum of drift to the yard. A single part halliard is quite enough, for it has only to lift the weight of the spar and sail. The main halliards are led through a block shackled to the jaws, and then pass through sheaves on each side of the mast. One fall can belay to the tabernacle, the other leads aft to the cockpit. Thus the sail can be lowered to a position of two reefs without leaving the cockpit. The boom can be bowsed down with a tackle to get a hard luff. The luff can either be laced to the mast or hoops can be used. Probably the lacing is good enough. The objection to this rig is that if the sail is cut as it must be with a long luff, there is no support to the yard when three reefs are tied in. Personally I would set a trysail sheeted to the boom itself, rather than put in the last reef. There would be considerable drift to leeward of the yard and there might be a lot of chafe.

An alternative to the gunter as shown is the true sliding-gunter. This rig might be better for a larger yacht, but for the small yacht under consideration the wire span is good enough. In the *Yachting World*, September, 1940, page 250, Conor O'Brien describes some very practical and neat fittings for a sliding-gunter, but they would have to be made by a craftsman and would be expensive. I have never seen a sliding-gunter, but before the advent of the Bermudan rig, the gunter-lug, fitted with a span, was the favourite and most successful rig in the small classes. The half-rater, the one-rater, the two-and-a-half-rater and even some of the five-raters were so rigged. Having a longer luff than the gaff rig it was bound to be better for racing, but the long yard precluded its use in the larger yachts. The gunter rig has more weight aloft than the Bermudan, and but for the question of mast lowering I would greatly prefer the Bermudan, perhaps with the luff rope of the sail running in a groove on the mast instead of a track. This is very neat, but it has the great disadvantage that the shrouds cannot embrace the mast or be shackled to a mast band. They have to be shackled to tangs screwed or bolted to the mast. I think that a loose-footed mainsail is better than the laced variety. This was introduced because the *America* had very flat sails and this was held to be the reason why she made an exhibition of our fleet. The real reason was that the *America* had a balanced hull with a perfect metacentric analysis. We now know that the aero-foil must approach the parabola. This is to some extent attained by a loose-footed sail. It is most important to be able to vary the strain on the foot of the sail, and when it is stowed it must always be quite loose. Again one cannot expect a sail to lower handsomely if the clew is hauled out bar taut. This is true of a gaff sail with hoops, but even more so with the Bermudan, whereas with a loose-footed sail it is easier to ease the clew before lowering the sail.

The measurement of sail area is a simple process. The headsails and the Bermudan mainsail are simple triangles. Take any side of the triangle and from the opposite angle drop a perpendicular. The area of the triangle is the base multiplied by half the perpendicular. Whereas our scale is half an inch to a foot we can use a scale of one inch to the foot and multiply the base or the perpendicular by two.

In the case of a gaff sail we divide the sail into two triangles. I found among some old drawings a sail-plan of an old-fashioned cutter, see Fig. 19. Probably I made it thirty years ago. On an inch scale the L.W.L. measures 16 ft., so the design, which I have called *The Ark*, affords a standard of comparison between a gaff rig and a gunter rig. To find the area of the jib, multiply IJ by OK, divide by 2 and the quotient is the area of the jib. The plan in Fig. 19 is drawn to a scale of $\frac{1}{4}$ in. to 1 ft. Either use this scale or make the appropriate correction. The same process gives the area of the staysail.

The mainsail is divided into two triangles by the line BD. Treat each triangle as before. Multiply BD by MC and AD by BL. If you employ the same inch-to-a-foot scale multiply by 16 and then add the two areas together.

We require the geometric centre of effort to work out the stability factor. It has no other value. In the case of a Bermudan rig the process is simple. Bisect two sides of the triangle and join the points of bisection to the opposite angles. The crossing point is the centre of area of the sail, or as some perhaps incorrectly term it the centre of gravity. Refer to Fig. 17. The centres of staysail and mainsail are shown. Join these two. At the centre of each sail raise a perpendicular and using any convenient scale cut off a length proportionate to the area of the other sail. Thus in the case of *Paida* the perpendicular from the centre of the mainsail is 47, the area of the foresail. That from the centre of the foresail is 130, the area of the mainsail. In this case I doubled the area, giving 260. I measured 2.6 in., and similarly for the other sail. Join the points 130

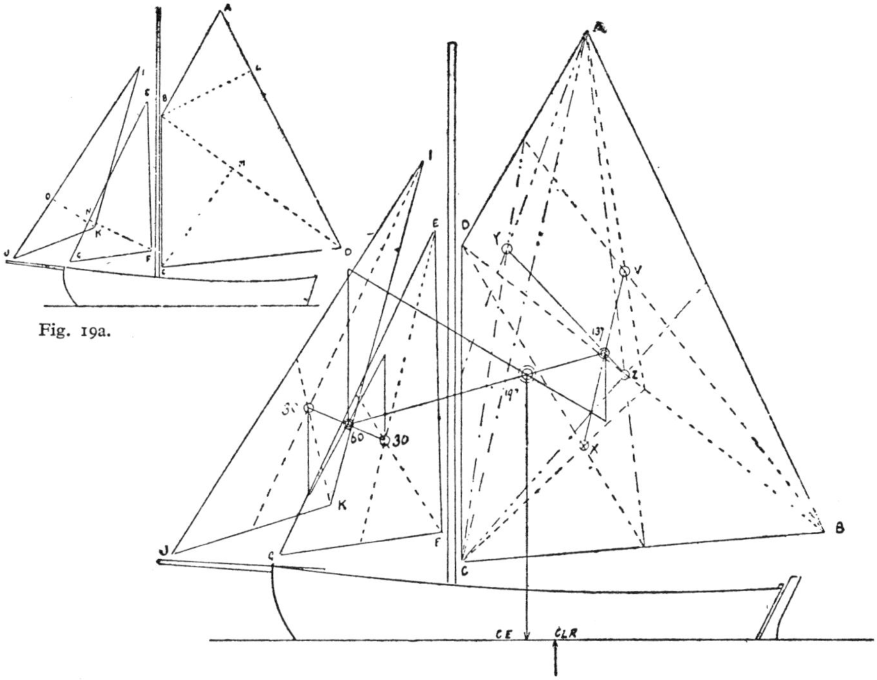

Fig. 19a.

Fig. 19. *ARK:* sail-plan of an old-fashioned cutter.

and 47 on the drawing and join the centres of the two sails. Where they cut each other is the centre of effort of the whole sail-plan. From this point drop a perpendicular and mark the point where it cuts the L.W.L.

I wish to make it quite clear that this is merely a conventional geometric centre of effort. It has no relationship whatsoever with the real dynamic centre of effort. Probably in a fresh wind this is close up to the mast. It is an unknown and constantly varying spot. When we come to the gaff cutter the problem is somewhat more complicated. Apply the same principles to jib and staysail and obtain a common centre of effort for the two.

In the case of *The Ark*, Fig. 19, each sail measures 30 sq. ft. Raise a perpendicular at the common centre, labelled 60 in the drawing. Divide the mainsail into two triangles by the line DB and find the centres of each labelled V and X and join them by the line VX. Now divide the mainsail into two other triangles and find their respective centres Y and Z. Join Y and Z. The point where the two lines VX and YZ cut each other is the centre of effort of the whole mainsail. This centre is labelled 137, the area of the sail. From the perpendicular rising from 60 cut off a part proportional to 137 and from that falling from 137 cut off a part proportional to the area of the two headsails. Join 60 to 137, and the spot 197 where the two lines cut is the centre of effort of the total sail-plan. Drop a perpendicular as before and mark the locus on the L.W.L. This finishes the measurement of the sails.

We have now to determine another mythical locus, the centre of lateral resistance, in the geometric sense alone, for this centre like the C.E. is an *ignis fatuus*. It wanders along the L.W.L. with the varying speed of the vessel. To find the C.L.R. cut out a silhouette of the underwater profile as in Fig. 18 taken from *Paida*, marking one of the sections, preferably the central section, in this case No. 5. Poise this silhouette in the usual way and we obtain the C.L.R. It is senseless to waste time calculating this locus. In any case it has no virtue whatsoever. The distance between the two centres is known as the "lead." Again this is a mere geometrical conception and has no practical value. Referring to my own yacht *Vindilis*, were there any real value in leads, stowing the mizzen ought to make an enormous difference in her balance. In actual fact it does not. When we took Mr. Crossley's *Edith Rose*, a metacentric yacht, out for her maiden sail, we first set the main and staysail. Under this combination she sailed herself to windward. We then set the big jib on the bowsprit, which would make a very large alteration on paper to the position of the C.E. But in practice it made no difference at all, the yacht, still unattended, made a perfect course to windward.

I have often asked experienced yachtsmen what happens if the mainsail is lowered and the yacht left to herself with her headsails set. They nearly all say she will fall off to leeward, and eventually find a course down wind. In actual fact the exact opposite happens. In spite of the enormous lead the yacht come up to windward and her course varies with her type. My *Sandook*, a Plymouth Hooker, came up to a broad reach and there she remained and would sail herself thus all day. Probably the stronger the wind the more she would come up. *Memory*, a little 18-ft. L.W.L. yacht that Mr. Suffling built to one of my early designs, with jib alone, and that on a short bowsprit, would come up to the wind and maintain a course well to windward. Still smarter craft will come up and go about with headsails alone. Where are your centres? The fact is that if the hull is unbalanced no disposition of sail will maintain a balance. If the hull is balanced the yacht will have a very wide tolerance, and the sails can be placed more or less as you wish. It is universally recognized that the Bermudan rig shows up bad balance and that the lead for this rig must be at least twice what it should be for a gaff rig. Many who have converted old-fashioned gaff-rigged cutters to Bermudan have discovered this fact by very unhappy experiences. I would go as far as to say that it is safer to adopt the gaff rig for an unbalanced hull.

I must apologize to the learned for my very simple explanations, but there may be some who do not understand geometrical problems and they may find my nursery teaching useful.

You will naturally want to ask me how to estimate the amount of sail a design will carry comfortably. Here again Admiral Alfred Turner comes to our rescue with one of his simple formulae:

$$\frac{\text{Ballast in tons} \times \text{distance in feet of centre of gravity of ballast below the metacentre}}{\text{Sail area in square feet} \times \text{height of the centre of effort in feet above the metacentre}}$$

This factor ranges from $\frac{16}{10{,}000}$ to $\frac{24}{10{,}000}$ and the 10,000 figure is neglected for convenience, leaving the factor at 16 to 24 for sea-going cruisers. In *Sinah* it works out at 17, and *Paida* is not so good with only 16.

This formula shows graphically how essential it is to keep all hull weights as low as possible and get as much ballast on the keel and as low down as we can. Here again the shellback says, "I do not want speed, and I want a strong ship." The idea is not so much to gain speed but, as I have said above, to obtain weatherliness. Recollect that the old Hereshoff 6-metre yacht *Suilven*, caught outside the Western Hebrides, was able to beat to windward under her trysail, weather the point, and run for safety. There was a heavy wind and a big sea. We shall come to the question of weight in construction later on.

I realize that this chapter has wandered from the subject of actual design, but it is useful to have some knowledge of the reasons for things that we do. The problems of balance, centres and leads can be worked out with models. The heated discussions on the metacentric system of balancing gets us nowhere; the whole problem should be treated practically. Either these theories are true or they are not, and *I* feel experiments with models, and if possible with real ships, will give the solution. I have been asked, "Can you explain why a good shelf is associated with good balance?" I do not know, and at the moment, judging by the discussions, no one does. But I do know that all my metacentric designs have produced balanced hulls, and with Bermudan rig none of my earlier designs have done this.

CHAPTER VII

Rigging and Gear

THE lowering mast is of immense antiquity, older than the invention of the anchor, in Europe at any rate. Thus in the *Iliad*, Book 1, Section 430, Homer tells us that: "When they arrived at the deep harbour, they furled the sail and stowed it in the black ship, and the mast they lowered by the forestay handsomely and eased it down into the crutch. They then rowed her with oars to her berth, hove out the mooring stones, made fast the sternfasts, and landed on the shore." It is thus obvious that the sailors of Agamemnon's hollow ship had no anchors but used large stones.

The lowering mast would at first sight appear to lack strength, and to depend entirely upon its rigging for support. This may or may not be true according to the type employed. It is used in the Thames barge, the Humber keel, in a vast number of fishing boats, and in many of the sailing craft found in distant seas. It is therefore obvious that if properly designed it is a practical contrivance. It may be housed in an orthodox tabernacle as found in the yachts that are built on the Norfolk rivers. This construction is shown in Figs. 20 and 21. Two cheeks of wood form the sides of the tabernacle, which is completed by another piece forming the back. The cheeks are bolted to the keel and carefully kneed to the mast-beam. A bolt at the top of the cheeks forms a pivot for the mast. The hole in the mast must have a metal bush, and both bolt and bush must be adequate to take the downward thrust of the mast when the rigging is set up. On the Norfolk rivers it is usual to balance the weight of the mast as shown in Fig. 20.

In the Thames barges and the Humber keels the mast is not rigged in a tabernacle but in a lutchet. This is constructionally the same as a tabernacle, but the mast heel does not go below deck. If it does a special hatch has to be made for it, and it is difficult to keep this watertight. Naturally, unless the lutchet is high-sided the stability of the mast depends largely upon its rigging. On the other hand the presence of a long mast hatch weakens the foredeck and demands extra stringers and knees. I may be wrong in my distinction between a tabernacle and a lutchet; it is possible that the same thing has a different name in different localities.

In the case of powered fishing craft the lutchet is generally a steel structure bolted to a stout deck beam and the down-thrust may or may

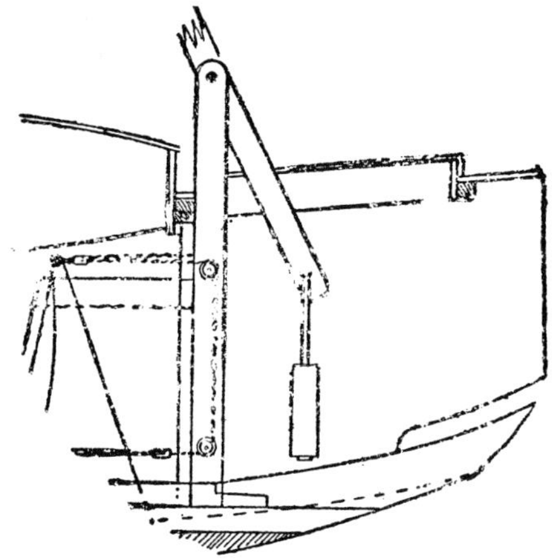

Fig. 20 (on right). A balanced mast in an orthodox tabernacle, as used in yachts on the Norfolk rivers.

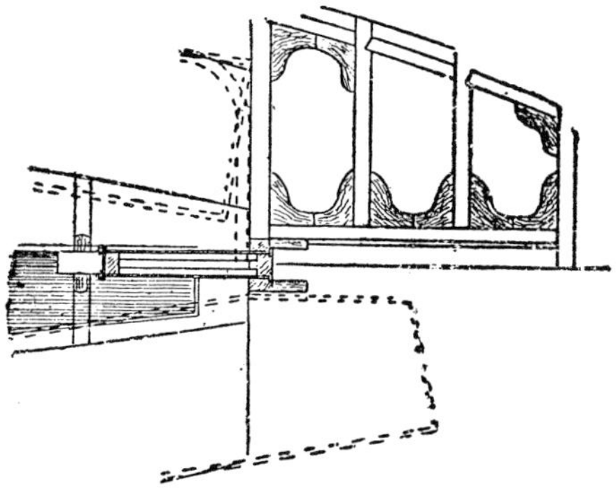

Fig. 21. The deck construction in the way of the mast well. The forward beam will be an extra strong one, and the carlines, or fore and afters, will be well kneed as shown to maintain the solidity of the deck, which is weakened by the well destroying the continuity of the beams. The hatch must be bedded down upon draught excluder to render it watertight, and a rubber pad must be fitted at the mast end of the well.

not be taken by a stanchion. This is also the method adopted in the Z 4-tonners. A steel lutchet is bolted to a deck beam and is supported by two stanchions. The whole thing is delightfully neat and effective. Here again the mast depends almost entirely upon the rigging. If this lets go, as the Americans say, over goes the mast. The last method, the principle carried to its logical conclusion, is found in modern racing dinghies and in the Scandinavian Skerry cruisers,the tooth-pick class. Here the heel of the mast is just shipped into a shallow cup, and its security depends entirely upon the rigging. In my humble opinion this is not ship-shape and Bristol fashion for a yacht, but all right for a racing dinghy, in fact good, for if the rigging goes the mast does not break, but just topples over the side and a new shroud puts things right again without expense or material loss of time.

A lowering mast has many advantages in a small yacht and I have figured it in *Paida*, Fig. 17, page 69. Not least of these advantages is that the mast can be lowered for attention to the gear at the mast head. For a sea-going yacht, however small, that is to be employed solely for waters where a lowering mast is not necessary, I would always prefer the orthodox mast stepped on to the keel. Now if this is wedged tightly so that there is no play at the partners, the mechanical conditions present are identical with those found when the mast is in a lutchet. The strain is localized at one spot, at deck level. The mast should, as is the custom in modern racing craft, be fitted with a rubber collar instead of with wedges. A steel band must be shrunk over the lower end of the heel otherwise the mast may split at the tenon, which is a weak place; when planning the keel construction in the way of the mast, do not overlook the fact that there is a tremendous downward thrust at this spot. In the case of some of the ocean racing class this amounts to several tons, and some of them developed leaks at the scarph between keel and stem.

The old-fashioned type of yachtsman will say, "Why have these strains?" The answer is that unless the luffs of the headsails are set up bar-taut the yacht will not go to windward properly. One day I was stretching a new suit of sails on *Vindilis* and they were all hanging in bags. A much smaller yacht, *Papoose*, was rapidly overhauling us, but as soon as I had taken a general pull on the halliards *Vindilis* at once held her place. Again, if the luffs are not well set up and the mainsail is not hauled out on the foot, the yacht may carry strong weather helm which at once disappears when the sails are properly set. The headsails cannot be set up unless the preventer-backstay and the runner are both taut, and the combined stress of the rigging aft and ahead of the mast causes a big downward thrust.

Any weight we can save in mast and gear adds largely to stability. In the case of *Sinah*, Fig. 22, I calculate that if we can save twenty pounds in the weight of the mast it is equal to a hundredweight of lead keel. This simple fact gives cause for deep thought! I have read that Dixon Kemp calculated that 25 per cent. of the heeling moment of a sailing ship was due to the weight of her mast, gear and sails. This may be an exaggeration, but it is a fact that these weights have a very great influence in diminishing the stability of a yacht, and it is obvious that especially in small craft weights above deck must be cut to the bone. Therefore a hollow mast is highly desirable in these craft. They are now so well made with greatly improved glue that they can be regarded as perfectly trustworthy. If you can obtain a silver spruce mast, and this is possible only in the small classes, it may take the place of a hollow mast if there is any material saving in price. I would shun a hollow boom, for the moment

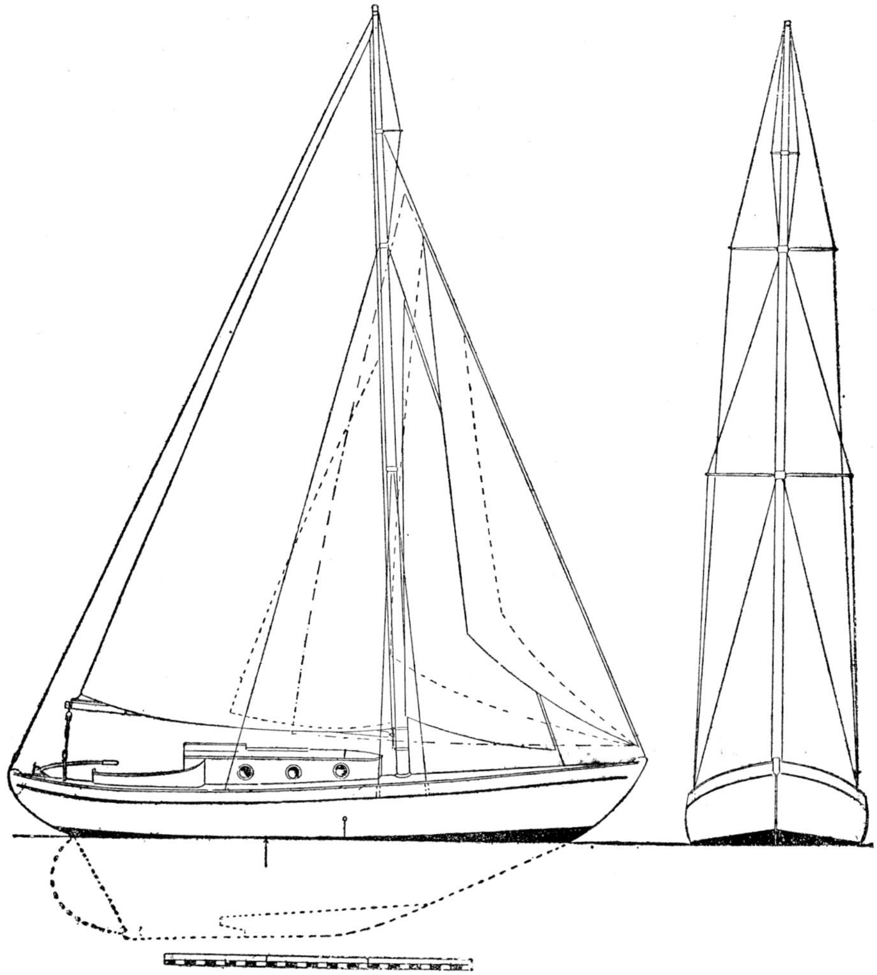

Fig. 22. *SINAH:* sail plan.

a tube is bent it loses its strength. I often wonder whether Nutting lost his ship with all hands because the *Leif Ericsson* had a hollow boom.

In deciding upon the scantlings of the spars, if a spar can have a scantling, and the size of the wire used for the rigging, one can safely follow racing practice. The strains thrown upon the spars and rigging of a racing yacht are in general higher than those present in a cruiser. but the standard of comparison must be displacement not linear dimensions. The weight aloft found in the majority of cruisers can safely be reduced and increased stability gained. You will note that here and always I am insisting upon stability. Never forget that stability varies as the fourth power of the linear dimensions. We cannot afford one pound of unnecessary weight in a small cruiser if we are to do good work to windward, and remember that ability to go to windward may mean safety in harbour instead of danger outside.

Some will say, "Why all this talk about windward work?" Chapelle on page 87 of *Yacht Designing and Planning* says, "The advent of auxiliary engines has enabled yachts to avoid beating." Perish the thought! The engine is good in its place, in fact in many cases essential, but if it is to avoid beating, then away with it. Or be logical and design a "fifty-fifty," a "motor sailer." Not only is going to windward the salt of sailing, the highest pleasure that sailing affords, but one cannot trust one's ship and one's life to a motor, for to drive a yacht to windward in really bad weather it must be very powerful, and heavy enough to kill the performance of a yacht under sail, and by its size curtail the living room in small craft. Such engines are also expensive to run. I hold a very strong opinion that when one is compelled by imminent danger to get under way and make an offing, it is essential to possess a craft that is able by virtue of good design, strong but not heavy construction, and sound gear, to beat her way out to safety under sail, and I should feel unhappy had I to trust my life to any small marine engine.

In my opinion the Bermudan rig is the best for yachts under 10 tons or thereabouts. It is simple and efficient, and saves the maximum of weight aloft. Being something new it is naturally opposed by the older type of yachtsman. We are told that it has inevitably brought in its wake an elaborate and highly stressed form of rigging and masting. This may be true in the metre and the ocean racing classes, but for an ordinary cruiser elaborate rigging is quite unnecessary. In my design *Omicron*, published in the *Yachting Monthly* for August, 1940, I drew a rigging plan with shrouds running to the mast head spliced over the hounds in the ordinary way. *Thuella* also shows the same arrangement, see page 144. I adopted this rig in my X-boat type of yacht, *Moyezerka*. She had the ordinary mast hoops and all worked in a most satisfactory fashion. One pair of cross-trees is generally sufficient in any cruiser. It is, of course, true that if for racing purposes one wishes to cut down weights almost to the limit, and often beyond the limit of safety, then complicated rigging is necessary. Personally I would have the mast a little stouter with simpler rigging, well knowing that I am still saving weight on the gaff rig. The weight of the gaff and of the blocks attached to it high up has a most

Fig. 23

VINDILIS

Mainsail	.. ..	230 sq. ft.
Mizzen	.. ..	47 sq. ft.
Foresail	.. ..	70 sq. ft.
Jib ..	.. ..	63 sq. ft.
T.S.A.	.. ..	410 sq. ft.

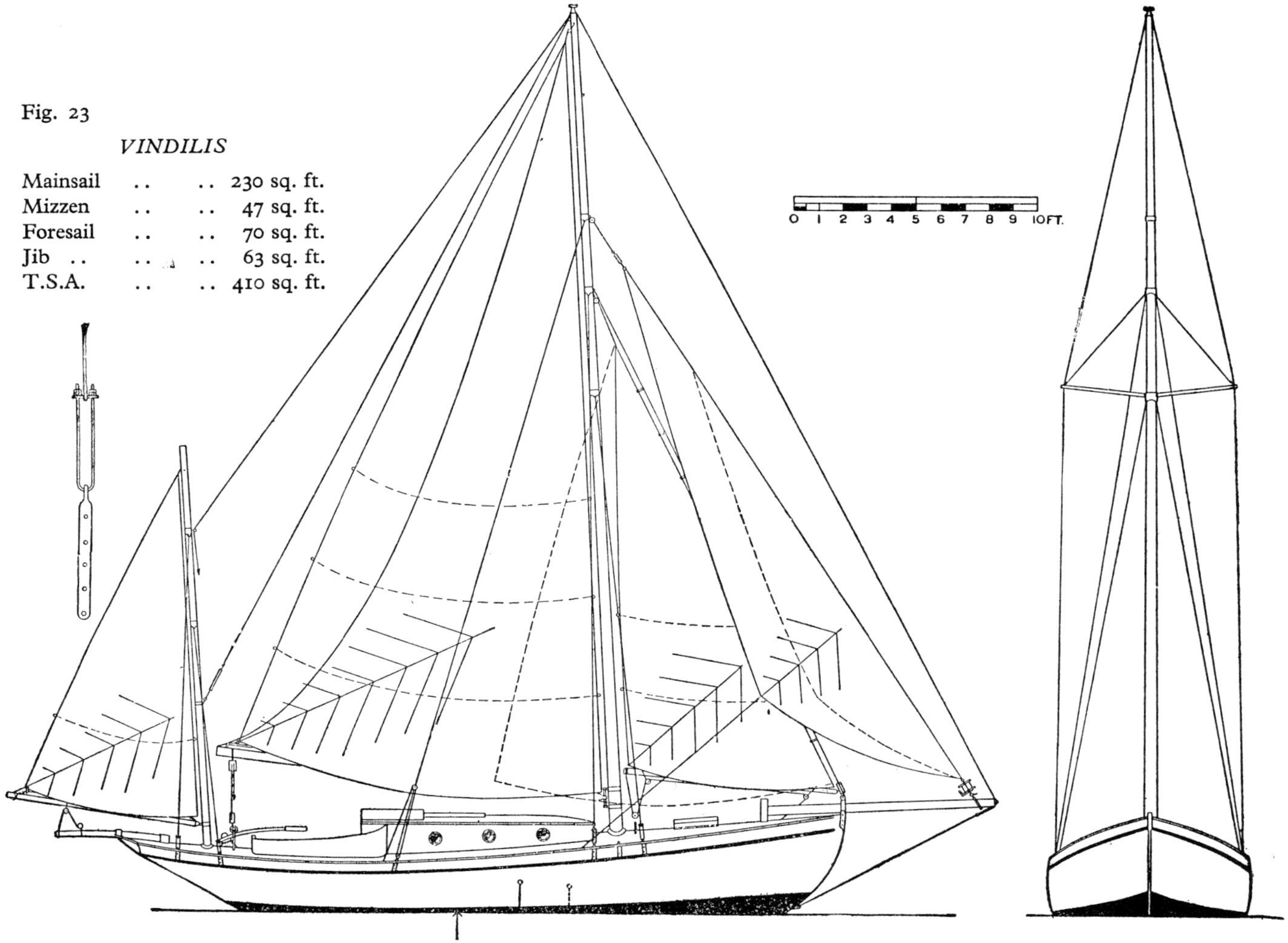

PLATE 5

(left) *Seasalter* (30′ LWL), Aristene design. Built 1937 by Peter Clausen of Port Adelaide, Australia for the Rev. Guy Pentreath. After extensive cruising in Australian/Tasmanian waters, she is now in New Zealand. *(Photo courtesy of Michael Pentreath).*

(right) *Quest* (20′ LWL), Vindilis I design. Built 1937 in Auckland, New Zealand for L.R. Wood. Now in Australia, *Quest of Sydney* does well in local races. *Dilys*, her English sister ship was built in 1936 by Vernon E.B. Nicholson for himself and is now in Orkney.

PLATE 6 (above) *Erla* (24′ LWL), Sinah design (THB's first 'metacentric' design). Built 1938 by Viggo Hansen, Copenhagen, for E.F. Hingeley. Sailed to Britain and subsequently to British Columbia. *Amiri* was built/owned by John Hartley, Victoria, Australia in the 1970s and another is believed to be building in Queensland.

(below) *Tramontana*, Cayuca design. Built (as *Morena*) 1934 by Anderson Rigden & Perkins, Whitstable, for Cmdr A.R. J. Southby. Now Falmouth-based, she is the largest known HB boat in the UK and, perhaps, anywhere. *(Photo courtesy of Michael Wilson.)*

detrimental effect upon stability, and so the rig is unsuitable in a small yacht where every ounce counts.

As regards stresses; it is as essential with a gaff rig as with any other to set up the headsails bar-taut to get the best windward performance, and there is no greater stress in an ordinary Bermudan cruiser than in a gaff cruiser. My *Vindilis* had quite simple rigging, and one pair of spreaders, and she has never shown the slightest sign of any weakness aloft. Probably she is too heavily rigged. The stays are set up with ordinary lanyards and the runners have the orthodox tackles. It has been said that the factor of safety in a complete Bermudan mast and rigging is of necessity lower than that of a normal gaff rig of the same sail area. I absolutely disagree with this statement. The factor of safety in a gaff rig is complicated by the weakness of the jaws and of the parrel. How often do we read of the gaff jaws carrying away? And the more so because the Bermudan rig can be 20 per cent. less in area for equal efficiency on all points of sailing except on a very broad reach and running. The statement that a Bermudan rigged yacht needs lighter construction to obtain adequate stiffness, is, I feel sure, wrong. The weights aloft area for area are less in the case of the Bermudan, and, as I have shown, weight aloft kills stability. It is as essential to obtain the maximum amount of stability in a gaff yacht as with any other rig. The centre of gravity of a gaff rig with a topsail is as high as that of a Bermudan; every fisherman knows that it is the topsail that gets the boat along, and the Thames barge skipper hangs on to his topsail with a brailed-up mainsail. Then we are told that "all that we have gained by the Bermudan rig and light construction is *speed*." Really we have obtained something far more valuable, *ability to go to windward*. We should design sailing yachts to go to windward under sail and not to be driven by petrol. If our yacht will point high and fore-reach, we have gained speed to windward without undue speed through the water. A short examination with some paper, a scale, and a protractor, will graphically demonstrate this, or it can be worked out with traverse tables. From my own experience with *Sandook*, a gaff cruiser, and *Vindilis*, a Bermudan of about similar size, I know that there is no comparison between the two rigs. The Bermudan is simpler, safer to handle, and far more effective.

After this digression let us return to the question of standing and running rigging. Let us keep the standing rigging as simple as we can, and as light as possible consistent with sufficient strength. All shrouds and stays must fulfil a definite purpose and do it to the best advantage. Every stay and shroud must have an antagonist. The wider the spread of a stay or shroud the greater its advantage or effectiveness, in fact I think that I am right in saying that the tension varies as the cosine of the angle of spread. Thus the greater the beam the less the tension in the wire, allowing lighter stuff to be employed. The value of channels in a narrow vessel is therefore obvious. The lower the shrouds the wider the angle between them and the mast, the greater the advantage at which they work, and again the lighter need the wire be. This is why the modern yacht has her rigging low down, as seen in the sail-plan of *Sinah*. A study of the sail-plan of *Vindilis*, which has in practice worked excellently, shows a different arrangement. The cross-trees are placed at the middle of the mast. The Bermudan sail tends to bend the mast aft near its central section, so the lower shroud is shackled to the same band that carries the spreaders, and is brought down to the for'ard chain-plate. The pull of this shroud counteracts the backward pull of the central part of the mast. The higher shroud is shackled to the same band that carries the forestay and the fore-halliard, so that the backward pull of the shroud balances the forward pull of the forestay and staysail. The jib-halliard and the runner are attached to the same band, again balancing each other. Finally the topmast forestay is counterbalanced by the preventer backstay shackled to a band on the mizzen mast, and the pull of this stay is transferred to the deck by the after mizzen shroud which shares the same band. The lower mizzen shroud leads for'ard to counter the back pull of the sail. A study of these wires will show that each balances and supports the other, there is nothing redundant. The bob-stay counters the pull of the forestay and of the jib, and there are also bowsprit shrouds which are useful to stand on and to catch the flukes of the anchor. Two centres of effort are shown, the dotted line marks the position on the L.W.L. of the centre of effort when the mizzen is furled. We hand the mizzen in strong leading winds and one would anticipate lee helm, but as a matter of fact under these conditions *Vindilis* carries considerable weather helm. This shows how ridiculous it is to pay any attention to these mythical centres. Actually the sail-plan shown is not that of *Vindilis*, but of *Lindy II*, a sister ship built to the same moulds. The bowsprit is rather longer and the mizzen smaller, and the mizzen cross-trees have been omitted. They give a certain big-ship character to *Vindilis*, but are weight in the wrong place and unnecessary.

We now come to running gear. This should be all of flexible wire running over sheaves amply large to prevent a sharp turn in the block. The tackles and falls should be of Italian hemp. *Vindilis* came out with manilla but gradually replaced it. Italian hemp lasts so much longer than manilla that it is in the long run not much more expensive. It is far nicer to handle and is stronger.

The main halliard can be rigged in two ways. The wire can run over a sheave in the mast, and the fall be fitted with a block through which is rove the tackle. This is brought down to the deck on each side and gives a two to one increase of power. This is not enough and a purchase is necessary. A luff-tackle purchase may be fitted to the end of one of the falls, giving a six to one advantage. I have this arrangement on *Vindilis*, but it would be far better to purchase the tack down with a luff tackle or use a small winch. Nothing elaborate is necessary. The sort of thing one sees on fencing wire would do quite well. The goose-neck is arranged so that the boom can rise and fall.

Another plan, not so neat but safer, is to have a block on the head of the sail, and a block or cheek sheave on each side of the mast. This gives the same double purchase, but if one side fouls the other still functions. If the main halliard gets off its sheave and jams, then the sail cannot be lowered and one is indeed in trouble. If the former method is adopted the sheave must be most carefully fitted and have a deep groove so that it is practically impossible for the wire to jam. The mouth of the mortice in the mast, I do not know its nautical name, must be lined with copper.

The jib halliards will be a single wire and a three to one luff-tackle purchase, or jig in American. If the ordinary double halliard is fitted it will roll up when a Wykeham Martin furling gear is used as it generally will be. It is one gadget that never lets one down if used properly. The staysail should be tacked down rather than purchased up. These arrangements keep all the gear on deck and avoid windage and weight aloft.

The sail-plan of our old friend *Sinah*, page 78, shows the modern all-inboard cutter rig. If the forward overhang is sufficient to dispense with a bowsprit, which apart from financial reasons it should be, this seems to me to be the handiest and most effective rig that has ever been devised for all but very small yachts, which would have a single headsail. The mast is most efficiently stayed, and the headsails are easily handled. If sail has to be reduced it is easy to lower either the jib or the staysail. Large, light sails can be employed, and being behind a stay and not aft of a mast they exert their full driving effort. There must be enough drift between the two stays to allow the jib to come over easily in stays. Many of the more modern rigs with both stays close to each other demand a trained crew to go forward and pass the jib over the forestay. There is a good reason for the absence of a topmast stay.

The "parrot perch" jumper stay strut and the double-span make the head of the mast into a kind of girder. The tension of the preventer backstay is transferred to this span and from it to the struts which exert a backward thrust countering the forward pull of the jib stay.

I am sure that it is a mistake to run the jib halliard up to the mast head as was done in some of the ocean racers, but not in Nicholson designed yachts.

I do not like diamond stays. In *Sinah* the topmast shrouds, or as some call them the backstays, are brought down to the deck where they can be constantly inspected and the correct tension maintained. I think that the orthodox rigging screw of the bottle type is a poor thing; weak, complicated and expensive. On *Vindilis* I have the ordinary strainer that is used on every telegraph pole. Each limb has two nuts, one a locking nut. Occasionally one goes round with a spanner and sets up the rigging. The outfit can be made by any blacksmith; it is simple, cheap and functions perfectly, though it is of course not "yachty." This form of rigging screw is illustrated on the sail-plan of *Vindilis*, page 80. This plan shows a staysail fitted with a boom, but is this a "good thing" or is it not? The disadvantages are that it is added weight; it is in the way on deck when the sail is furled; and most important, it allows of no overlap and so

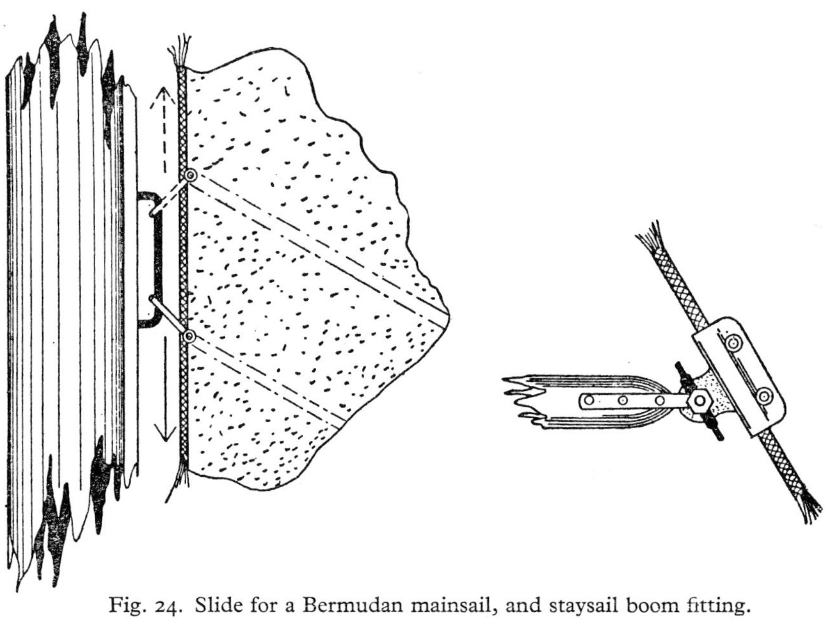

Fig. 24. Slide for a Bermudan mainsail, and staysail boom fitting.

does not give the mainsail that important funnel effect; in a word it is aerodynamically inefficient. It has, however, great advantages. In a short-handed ship it is very handy, for there is only the jib sheet to handle; it is a fine sail for running in a strong blow, under staysail, with or without a mizzen. If the sheet is eased off till the booms makes an angle of about 45 degrees with the mid-line of the ship, it boom out quietly, and the yacht runs easily and safely. Running dead under an ordinary staysail is a tiresome job. The sail flutters about and winds itself round the stay. Another real advantage is that the boom is a most valuable hand-hold when working on the foredeck. The boom swings from the forestay; the fitting we use is illustrated in Fig. 24. The sheet, which needs considerable purchase, must be so adjusted that the boom is not drawn either fore or aft, but hinges naturally without any tension on the forestay.

Runners even in quite small craft ought to be fitted with some form of Highfield lever.

The clew outhaul tackle must be so arranged that the fall makes fast on the boom close to the mast. It is then handy for constant alteration in the tension of the foot of the sail. Off the wind it can be eased off, but when the ship comes to wind the foot of the sail must naturally be hauled aft. When the mainsail is to be lowered the first thing naturally is to set up the topping lift; we all do this, but how many do the next thing, ease off the clew tackle? As I have already said no sail will lower handsomely if the foot of the sail is tight. Ease off the clew and use its fall as one of the tyers to make up the bunt of the sail. If this is done one will not forget this important duty before the sail is made up, and it will not be pulled out of shape by moisture.

If the slides of a Bermudan mainsail are to travel sweetly there must be no tilting or each will exert a braking action, and the accumulated resistance will be great. The slide shown in Fig. 24 has a long eye and the sail is shackled to this. When *Vindilis* came out the slides were bent to the sail with marlin. We had a lot of trouble till this was replaced with shackles.

CHAPTER VIII

The Lay-out

THE lay-out includes not only the cabin plans, but the general arrangements below deck level. Proper stowage of gear is as essential, if not more so, as human comfort below. Many a prospective owner has written to me about a design, and has told me that, "I want only the lines, sail-plan, and construction plans; I can manage the lay-out myself." In the case of a competent sailor this may be, in part, true; but, generally speaking, a designer who has thought out these plans for years and has seen his ideas blossom into yachts, can generally advise even hardened cruising men about the best arrangements for a small yacht. In one case, an experienced yachtsman built one of my 7-tonners. He moved a bulkhead a foot forward, but forgot the mast. When this was stepped the forecastle could be entered only by hard squeezing, and a really big man could not enter at all.

When we lie in harbour one of the greatest pleasures is visiting the yachts moored near by. Some are marvels of neatness and, even when they have just returned from a long cruise, they look as though they had just left the launching ways. At one of the meets of the Royal Cruising Club I visited the 6-tonner owned by the vice-commodore. She had just returned from a single-handed cruise to the Baltic, and was a marvel of order: there was not a bit of loose gear to be seen, the cabin was not only neat but polished and shining; all was a delight to the eye. The whole yacht inside and out looked like a bit of choice Chippendale, and this at the end of a prolonged cruise. On the other hand, I have visited small yachts which were mere rubbish heaps. In the one case there was a place for everything and everything in its place, in the other a place for nothing and everything everywhere, the whole a scene of desolation accompanied by a fusty smell.

In planning the lay-out we must first keep the essentials before us; in the very small sizes we can get little more than these bare necessities, and I would divide the essentials into two groups: those vital to the proper navigation of the ship, and those which are personal requirements. There must be a proper place or locker for the sails, warps, lead-line, watch-buoy, fenders, fog-horn, flares, and for charts and instruments of navigation.

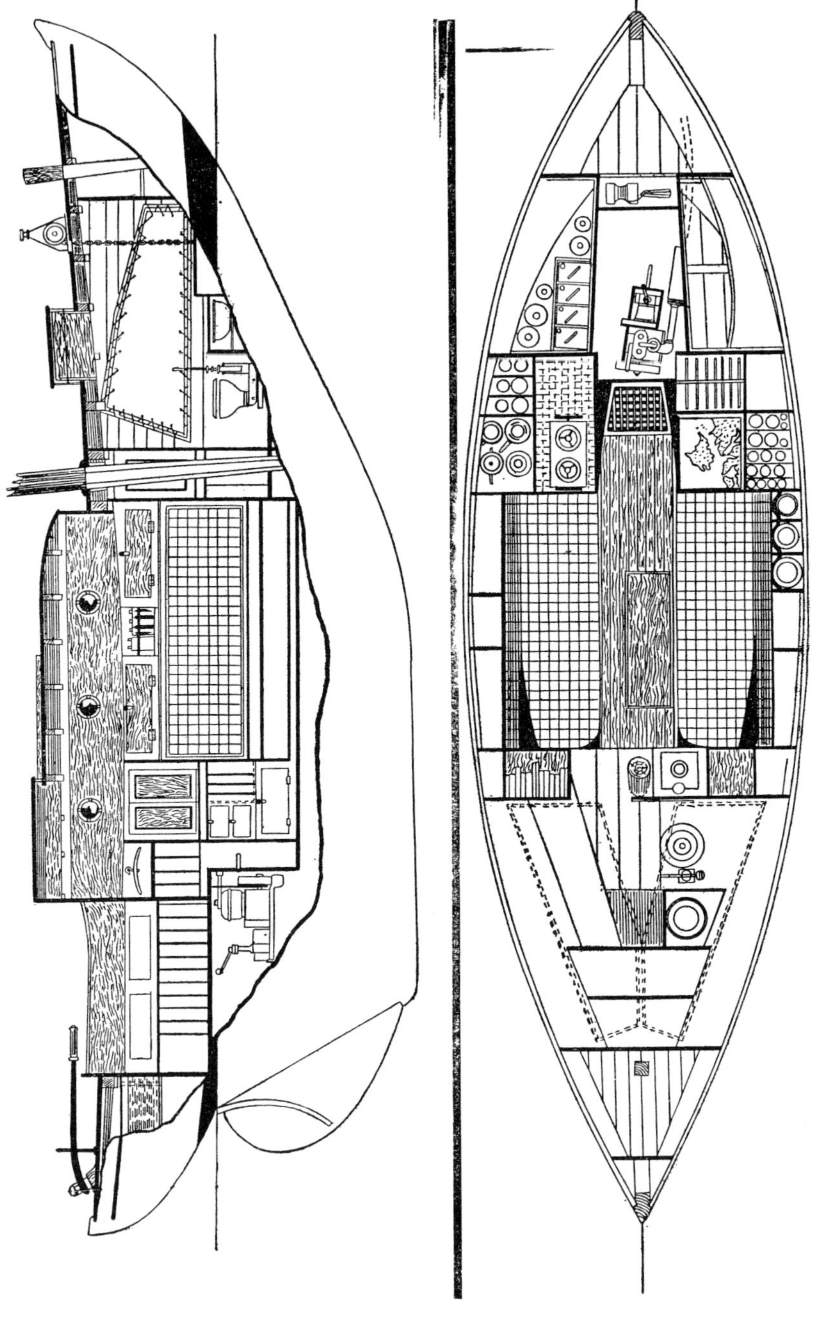

Fig. 25. *SINAH*: accommodation plans.

I always try to arrange for a chart-table or for a chart-box which forms part of the back of a seat, and can be lowered when necessary. I have tried laying out courses on the cabin seats, and I assure you that nothing is more conducive to sea-sickness. I am visualizing a yacht that really cruises, and not a mere coastal potterer. All the spare gear must be instantly accessible, warps ready to hand, and sails stowed dry in their labelled bags. All this can be arranged for in the smallest yacht, even one like *Paida*. In the very small sizes the forecastle is a huge locker, and it is easy to plan a manger for the sails, and hooks for the warps.

Cabin arrangements are not nearly as important as due provision for the orderly stowage of gear. The Greeks knew this well. When Xenophon, under the pseudonym of Ischomachus, asked his wife Philesia for an article in the house and she was not able to find it, he described to her a visit to a Phoenician merchant ship, and the great impression that it made upon him. "I have never seen gear so well arranged, or so many coils of rope and tackles stowed so neatly. A ship needs a large number of spars and warps when she enters port or puts to sea; much rigging when under sail, and contrivances to protect her against enemy craft. She carries a stand of arms for the crew, and each mess needs a set of household utensils. In addition, she carries a cargo which the captain sells for profit. All the gear necessary for these several functions was contained in a small store not more than fifteen by twelve feet. I notice that each article was so neatly stowed that it was ready to hand; it had not to be searched for, and there was nothing to cast off and cause delay when anything was needed in a hurry. I found that the bosun knew each particular locker so well that he could, even when on deck, say exactly where anything was stowed and how much there was of it. I saw this man in his off-duty time carefully inspecting all the stores most likely to be needed. I asked him why he did this. 'Sir,' he said, 'I am looking to see that all the gear is properly stowed, nothing foul, nothing missing. For when God sends a storm at sea there is no time for searching for gear or clearing it if foul. God threatens and punishes careless sailors and you are lucky if you escape with your lives. You are fortunate if, even when you show good seamanship, He brings you safe into port.' "

Xenophon then pointed the moral to poor patient Philesia: "If men tossing on the sea have to stow their stuff in a small space and can at once find what they want, how much more ought you, living in a large house founded on solid earth, furnished with ample store-rooms, be able to put your hand on what you need."

We must keep in mind the words of this Phoenician bosun, and see

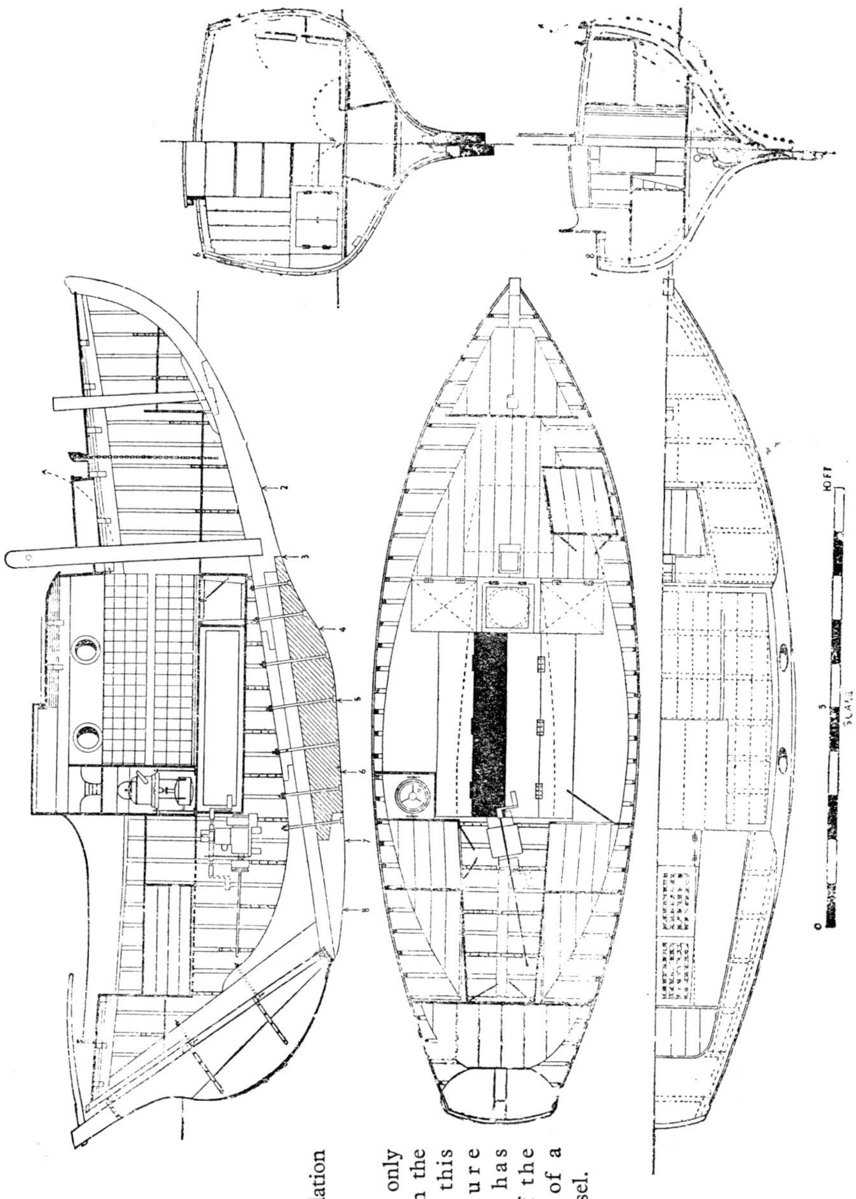

Fig. 26.

PAIDA: accommodation plans.

Although only 16 feet on the L.W.L., this miniature cruiser has many of the features of a larger vessel.

that we have a proper place for every bit of gear, and for every instrument necessary for efficient seamanship. Later on, and this has nothing to do with yachting, poor Philesia turned up in high heels, with her face painted; another long sermon, and due penitence. There is nothing new under the sun.

The cabin lay-out can be reduced to general principles. Certain things are indispensable, absolutely necessary, and must receive first attention. There must be comfortable, dry sleeping accommodation for every member of the crew. Men go to sea for a holiday, for rest and recuperation, and not to suffer unnecessary hardship. One, at least, of the bunks must be suitable for sleeping in under-way. In *Vindilis* I have canvas "lee-boards" which are lashed up to the hand-rail and make it impossible for the sleeper to be thrown off his bunk. The bunks can be of the usual iron-framed cot variety, they can be of the so-called "Root" type, or one can sleep on the cabin seat. If this is two feet wide this is probably the best solution. The cushion should be either Dunlopillo or the coiled spring type, I prefer to sleep on the seat because one has valuable lockers above the back of the seat. This turns down to open a large locker for the bed clothes.

The second essential is adequate cooking gear. This is hardly the time to discuss the best position of the galley, for it depends upon the size of the yacht, the way she is to be used, and the personal feeling of the owner, or rather of his wife, who will veto cooking in the forecastle because all the time she is cooking she is cut off from the social amenities of the saloon. This means it must be in the saloon, either fore or aft. In the case of a baby yacht, the galley will be a single gimballed Primus.

The third essential is sanitary accommodation. This is a most important question which, perhaps for reasons of modesty, is generally most inadequately dealt with. Accommodation is necessary even in small day boats. I once nearly bought a small clinker-built quay punt, designed by Harley Mead and built in his Cornubia Yard at Cowes. She was, I think, 18 ft. on the L.W.L., and whenever I went to Falmouth I admired her. However, I was not allowed to buy her, and had *Sandook* instead. The next time I saw her she had a small cabin. I said to the lady who owned her, "I see that you have added a cabin." "Of course I have," she said; "how can I take a mixed party afloat with no lavatory accommodation?" A very wise saying. I am inclined to think that the ability to lodge the sanitation in the forecastle, apart from the main cabin, decides what must be the smallest size of a cruising yacht. I am utterly opposed to a separate compartment in any yacht under about 12 tons. In the first place, it is absurd to sacrifice two and a half feet of valuable space in the best part of the hull for functions which are limited to a few minutes a day; in the second place, these small compartments, ill-ventilated, smelly and difficult to clean, have no advantages from the standpoint of privacy. A mere thickness of wood does not comprise seclusion, and for all practical purposes of concealment, apart from the visual, might not be

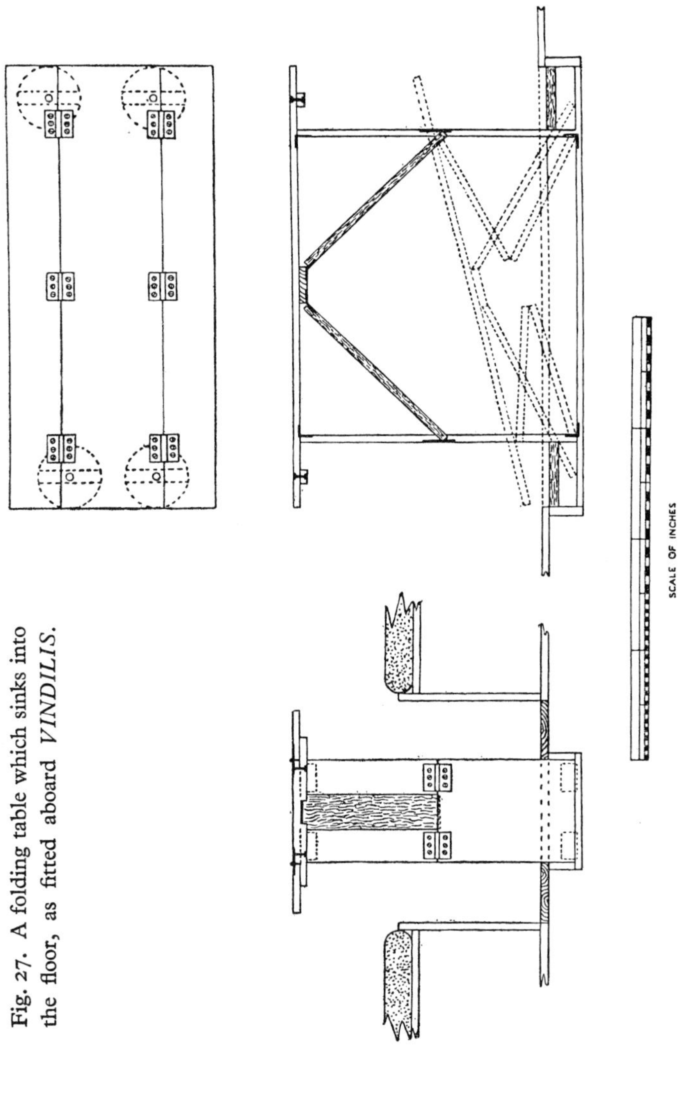

Fig. 27. A folding table which sinks into the floor, as fitted aboard *VINDILIS*.

there. Now if the sanitation is lodged in the forecastle, there is considerable secrecy, for one can enter the forecastle from the saloon for a variety of purposes. Never forget that, even when anchored head to wind, the current of air is from the stern forwards, and with an open forehatch, the use of the convenience is attended with no unpleasantness. Again, these contrivances have to be used at sea, when there may be considerable motion. An arrangement that, with skilled acrobatics, can be made to function in harbour may be quite useless at sea. With a mixed crew of four I have never, either in *Vindilis* or *Sandook*, found the forecastle lavatory any detriment, except once or twice at night. Under these circumstances, a bucket in the cockpit has sufficed. The underwater machines are not suitable for very small craft; they are too heavy, and too high. Nearly three feet sitting room must be allowed, but part of it can be gained by utilizing the extra height given by the forehatch. In such craft a bucket will be used. The compartment in which the bucket stands ought to be lined with lead or other metal, otherwise in time there will be a chronic smell, for with a wood lining adequate cleansing is impossible.

Fourthly, there must be comfortable sitting room for every member of the crew, and each position must be properly illuminated. The wooden seat should be about a foot high, the cushion makes it 14 inches. This is a most comfortable height, and allows for ample headroom. The minimum headroom is three feet over the top of the cushions. A back rest is essential. If the bed takes the form of a lowering cot, this provides the back; I think the motor people call it a "squab"; otherwise, one must be constructed. Of course, a small pad is more comfortable, but plain wood will do. A lamp at each end of the cabin will give light for four. Some may think that I am being too luxurious, but I am sure that comfort below is essential. When the day's sailing is done, dinner eaten and washing-up finished, then the crew settle down in the cabin to read, play bridge, converse, make up the next day's work, and so on. In harbour, one entertains friends from nearby yachts. Yes, there shall be comfortable sitting room, for during spring and autumn cruising a lot of time is spent in the cabin, and darkness falls early; hence the need for illumination.

Nothing of the nature of a Pullman berth can be allowed in a small yacht. They are far too heavy, and we have long ago agreed to cut down hull weight to the minimum.

A table is a necessity. In the small classes this can be attached to the mast and made to fold down out of the way when not in use. In the

6-tonners and over, I think that my folding table which sinks into the floor is excellent. I have now used it in *Vindilis* for some years and find it good (see Fig. 27) but it must be well made of hard wood.

There must be as many lockers as possible. A food locker or two, and a clothes locker for each member of the crew, and each must be made to use his own locker and no other. Whenever possible, the doors of all lockers should open athwartships. Those that open fore and aft are a nuisance, for they are continually flying open and the contents are spilled on the cabin seats and sole. I constantly read of the condition in the cabins of yachts that have had a bit of a dusting: "The floor was covered with a horrible mixture of blacking, pots, haddocks, Swiss milk, paraffin, and face cream." This means only that the yacht was thoroughly badly arranged. Shelves along the top over the seats are useless and a constant source of untidiness; their place should be taken by lockers. In *Vindilis* we had thirty-one lockers and each had its use. The ship was always stowed ready for sea and there was nothing that could carry away and make a mess.

Fresh water can be carried in tanks or in special tins. The Gem tins are, in my opinion, too large and heavy, so I have had some made of galvanized iron exactly like a petrol tin; they hold two gallons and weigh twenty pounds. One in each hand is a comfortable weight, and they can be taken ashore for filling without difficulty. In a small yacht they are more suitable and far cheaper than a tank, but there must be a definite place for them where they are firmly chocked off. I lay it down as a law of the Medes and Persians that a yacht must be permanently stowed below ready for sea. One ought to be able to get the anchor and sail off, knowing that nothing can carry away below. On no account use ordinary petrol tins for water unless they are repainted distinctively. Once in *Sandook* I had been cleaning all the lamps and decided that a wash and shave was indicated. So I filled a kettle and put it on the Primus. When it began to steam I poured it into the basin and shaved with it, and a very nice shave it gave. Then I began to wash up some cups and only then discovered that I was not using water. After that I decided to repaint the water tins. Once my daughter filled the riding light with petrol, because in her yacht they carried paraffin and petrol in the same kind of tin. The resulting blaze fortunately had no worse result than the loss of the riding light, which had to be dumped overboard. The situation of tanks must be carefully chosen. They are not well placed by the side of the cockpit, because they alter the trim of the vessel when emptied. The best place in a 7-tonner is under the saloon seats. The two must be joined by a pipe with a cock, which is closed when sailing. An ordinary pump is better than a semi-rotary. This type, unless fitted with a non-return valve, often fails to suck. In a larger yacht the tank will be under the cabin sole.

If the yacht is to be used in the spring and late autumn, and most certainly if she is in commission during the winter, some sort of stove is necessary. There are many on the market and the owner can choose the one he prefers, but it is dangerous to use any form of stove without proper ventilation. An iron stove, when hot, becomes porous and carbon monoxide can escape into the cabin. This is a most dangerous gas; it has no smell and produces insensibility and death without any warning, but the cabin of a small yacht can become a death chamber without any stove. Once, during the last war, I was spending a few days' holiday aboard *Sandook* moored in the Hamble River. We had taken all precautions to ensure a reliable "black-out," covering the ports with brown paper and shutting up all the hatches. With my three children I was sitting in the cabin playing Nap, when the light went out. We took it outside and trimmed it, and again it went out. At last I realized that it would not burn because it was short of oxygen. Had we gone to bed with all closed up, there would have been no lamp to warn us and we should all have died most peacefully from "anoxaemia," and have known nothing about it.

Head-room in a small yacht can be attained by the ordinary cabin top or by building up the top-sides. I have come to the conclusion that the second plan is the best, but I do not like to carry the raised top-sides to the bow, for this is very ugly and suggests a motor-boat, and it also means increased weight and windage in the bow, where it is not wanted, for anchor work on a raised deck is awkward and dangerous. The central turret can be made to look well, and seems to be the best solution. In a 7-tonner we can get sitting room under the side decks. A narrow deckhouse gives six feet standing room and wide plank-ways. All the woodwork inside must be kept light, and here the new resin-bonded plywood can be used with great advantage.

In drawing the cabin-plan, a large number of transverse sections must be made. The whole thing is rather tedious, but if due care is not taken, floors, seats and other structures will be placed in impossible positions, and in places maybe outside the skin of the yacht.

Especial care must be taken with the engine installation, and before this is done comparison must be made with the construction plan. Most engine builders now furnish tracings of their engines made to varied scales. In making the drawing, which is generally on tracing paper, one of these drawings is slipped into position under the tracing paper, and the engine traced on the drawing in the intended position. This is done on the elevation of the plan, and the transverse section involved. With these tracings a vast amount of time is saved, and it is easy to see if we are placing the engine in an impossible position. The drawings of the cabin-plan of *Khamseen* (page 155) show quite clearly how the cross-sections are made. The plans were made ten years ago, long before the manufacturers thought of making tracings of their engines for the use of designers. I think that the Austin firm were the first to make this valuable advance.

The original *Khamseen* was built in Malaya and was put on to a steamer and carried to Port Said. There she was put into the water and sailed to England partly via the canal from Cette to Bordeaux. Another *Khamseen*, called *Dorothea*, was built at Whitstable. I have brought this design up to date on modern principles. The original lay-out was designed for ocean cruising, but has now been modified for ordinary home cruising. Her lines are an expansion of those of my design *Zyklon* that was used for the well-known Z 4-tonners. The bunk arrangement is ideal for a mixed party, for any combination of sex can be accommodated: two women in the cabin, a man in the forecastle and another aft; one woman for'ard, three men aft; or three women for'ard and one man aft. I consider that in any yacht that is used for serious cruising there must be adequate space for gear where it is instantly available, so keep the bosun of the Phoenician freighter in mind. It would be possible to arrange two berths aft, but then we lose the stowage for sails, warps, lamps, and what not. These are in practice wanted aft, for in bad weather the forehatch cannot be opened, nor is it convenient to stand on the forecastle and deal with gear at sea. It ought to be stowed aft where it is wanted.

CHAPTER IX

Engine Installation

In Homeric times an engine was quite unnecessary. The hero made friends with a goddess, took her with him, and she at once provided a leading-wind. Thus, we read in the *Odyssey II*, 415:

"They stowed all their gear in the well-found ship, under the orders of the beloved son of Odysseus. Then Telemachus went aboard, preceded by Pallas Athene, and both sat down in the stern of the ship while the crew ashore cast off the stern-fasts, boarded the ship, and sat down on the rowing benches. Then flashing-eyed Athene conjured up a fair wind that sang over the wine-dark main. Telemachus called his crew and bade them man the halliards and they sprang to attention. They raised the fir mast and set it in the hollow step. They made fast the forestays and hoisted the white sail with twisted thongs of ox-hide. So the wind bellied out the sail and the dark surge sang loudly about the stem of the ship as she sped over the waves making her passage. Then when all was made fast they produced bowls full of wine, and poured libations to the immortal gods that are eternal, and chiefly to Athene, the flashing-eyed daughter of Zeus. So, through the night till dawn, the ship cleft her way."

We cannot now obtain such service and have to be content with petrol. For twenty years I had no engine in *Sandook*, and we always managed to get somewhere, if not always just where we had planned. But now, I speak in the pre-war sense, all the harbours and rivers are so cluttered with craft that an engine has become a necessity. To those who are obliged to get home at a definite time it is essential, and it is a definite element of safety in small yachts. Often a long passage can be made in calm weather and smooth water, whereas the yacht without an engine is becalmed, and we all know that long calms can be rapidly succeeded by strong winds.

Let us at once decide whether we are aiming at a fifty-fifty or an auxiliary yacht. If the former, the yacht must be properly designed for this purpose. For ordinary purposes it is not necessary to try for a speed higher than the ordinary speed of the yacht under sail. In most cases five knots all-out is sufficient, and this can generally be obtained from one horse-power for each ton of displacement. I have no intention of discussing the varied types of engine or the details of installation, but

PLATE 7

Erla under sail, Denmark.

Zingara (25′ LWL), Askadil design. Built 1936 by A.H. Moody & Son, Swanwick Shore, Southampton, for Dr R.E.D. Cargill of Ottawa. Launched by EB (Mrs THB), she was shipped to Canada and cruised on Lake Ontario, then migrated to the USA and is now in Maine. *Askadil* has made several Atlantic crossings and she and *Alexa* (ex *Envys*) are Solent-based; *Wendy* sailed round the world; *Naida* has visited Turkey.

PLATE 8

(left) *Sabrina* (22′ 6″ LWL), Yonne design. Built 1935 by Clemens, Portsmouth for A. Meadows. Ian Howlett is her third owner in 59 years. A major refit is nearing completion in Burseldon. *Yonne* made a Caribbean cruise but is now being rejuvenated in Rye. *Yarinya* (ex *Pride II*) is Netherlands-based after some years spent in Sark and Guernsey; *Destina* (ex *Phoebe*) is now East Coast-based and has cruised in European waters including the Baltic. Where is *Mischief III*? *(Photo courtesy of Ian Howlett.)*

(below) *Grey Owl* (19′ LWL), Zyklon design. Built 1938 by Everson & Sons, Woodbridge. Her owner and his dog are in the photograph. She was last heard of in 1975 in Mallorca. Not a Z.4 tonner.

there are a few important factors that I should like to mention. The most important and the most neglected is fire danger. American engines are fitted with flame arrestors and methods of conveying carburettor drips back to the engine, but here, so far as I am aware, these vital matters have been neglected. It is a mystery why most yachts are not burnt out. If possible there should be no petrol on board, and in any yacht of 10 tons or over I would have a diesel or heavy-oil engine, but the ordinary diesel and semi-diesel are too heavy for small craft, and too expensive, and the same applies in lesser degree to paraffin engines. The perfect auxiliary engine has so far not been evolved for small yachts. It ought to have two cylinders arranged V-wise to fit the V-shape of the run of the ship, and be as light as possible, fast running with reduction gear.

Shall the engine be centrally installed or offset, and if offset, on which side? The central position of the propeller allows the engine to be fitted farther aft than when it is offset, but unless the propeller is small it may have an appalling effect upon the sailing of the vessel. One may almost say that if reduction gear is used the shaft must be offset. If central, the aperture must not be in the rudder, or only slightly encroach into it. The best arrangement is to have a split rudder stock and run the shaft through. Then it works clear of the disturbed water round the run, and cannot affect steering. Or it may run out by the side of the rudder stock through a specially thickened part of the stern post.

I had a most unfortunate experience with *Vindilis* due to her central installation and to the large aperture in the rudder. On her maiden trip going down the Hamble the engine stalled. We were leaving the river on the ebb with a foul wind. Only headsails and mizzen were set. I at once put up the helm, intending to put her before the wind, when we could have held her stationary with wind and tide opposing while we got the engine going. But she paid off so slowly that she struck a mooring chain and made a nasty dent in her brand new top-sides. I found that she was very slow in stays, and generally sluggish. Under engine she took strong helm, I forget whether port or starboard, and manoeuvred with difficulty. The reason for the large aperture in the rudder was that the stern-gland was at the end of a tube that projected some four inches beyond the stern post. I had a new type of stern-gland fitted and was able to fill in about a third of the opening in the rudder. The effect was astonishing. Under engine she then took no helm at all, making a straight course unattended and she answered to the helm with ease and surety. Under sail she was measured against yachts that had previously outsailed her, and was at least a knot faster, while her slowness in stays had almost disappeared. I could never have believed that an unsuitably designed installation could have had such a really diabolical effect.

If the shaft is offset, on which side ought it to be placed? The answer is obvious: the shaft must be on that side which makes the propeller bias antagonize the shaft bias. In most cases a propeller on the starboard side pushes the bow to port, and a propeller on the port side drives it to starboard. There is probably for each hull a shaft angle which neutralizes this bias, but it is generally too obtuse to be desirable. On the other hand, a right-handed propeller tends to wind the stern round to starboard and a left-handed one has the reverse action. If we stand at the stern of the ship and look forward, a right-handed propeller moves clockwise, a left-handed counter-clockwise. The lower blades of the screw move in denser water than the higher, and so produce a winding action. Consider the action of a surface propeller. The lower blades alone reach the water. It is obvious that this wheel will act like a paddle wheel and swing the stern sideways, to right or left according to the direction of rotation of the wheel. Thus, a right-handed wheel will paddle the stern round to starboard. A totally submerged screw will have the same effect in less degree, and will drive the stern to starboard and the bow to port. We have seen that a propeller on the port side will drive the bow to starboard, so this will antagonize the bias of a right-handed propeller. Therefore, if we use a right-handed wheel, as the Americans call it, the shaft must be on the port side, and *vice versa*.

In the Z4-tonners the shaft is on the wrong side, but the little ships have very small propellers and they run straight. But in another yacht with a side installation on the wrong side the effect is most detrimental. She is quite out of control till she gets good way on. The ordinary engine runs counter-clockwise when viewed from the stern and has a left-handed propeller; therefore, put the shaft on the starboard side. But if there is reducing gear this will reverse the direction of rotation, and the shaft will have a right-handed propeller and will be placed on the port side.

An offset wheel works in less disturbed water and gets a better run of water than one behind a stern post, and is more efficient. Opinions seem to differ about the comparative drag caused by an offset or central position. There is not much in it. I am inclined to think that if the displacement of the yacht is not too high with reference to her length of load water-line, it is better not to have reducing gear, except in the case of a small, high-speed engine. The smaller screw is not so efficient under power, but it is far less detrimental under sail, and after all the engine is only an auxiliary.

The central installation can be rendered more efficient if the after side of the stern post is well streamlined in the region of the aperture, and the fore edge of the hole in the rudder should be similarly treated. Finally, with regard to the engine. The makers rarely supply proper filtration. Insert an "Auto-klean" filter in the pipe-line and you will never have blocked jets.

CHAPTER X

Hull Balance

THE riddles of naval architecture and aerodynamics were considered many thousands of years ago. In the Book of the Proverbs of Solomon we find that Agur, the son of Jakeh, pondered on the mysteries of hull balance and flight. He said: "There are three things that are too wonderful for me, yea four which I know not: the way of an eagle in the air; the way of a serpent upon a rock; the way of a ship in the midst of the sea; and the way of a man with a maid." Today the problem is not completely solved; the way of a ship is subtle and imponderable. Hull balance is the Cinderella of naval architecture; like Cinderella it has been neglected, and yet, like her it is the most beautiful aspect of design. In most books on designing, even of recent date, it is not even mentioned, and the majority of fishing boats, trading craft and cruising yachts are unbalanced. I am using the word *balance* solely with reference to the forward motion of a yacht; the balanced yacht when heeled tends to hold a straight course, whereas the unbalanced hull has a bias to port or starboard.

The first question we have to study is: what are the causes of unbalance? Why do so many ships pull hard on their helm when heeled? The second, the natural sequence: how can we eliminate these causes?

The fundamental source of unbalance is the double wedge-form which for decades has dominated British, American and French design, but has not attracted the Scandinavians, the Mediterranean races, nor the Arabs; nor have the Chinese and other seafaring people adopted this curious model.

The double-wedge is shown at A in Fig. 28. This vessel has a water-line which is an isosceles triangle; the underwater profile is a right-angled triangle. With modifications this form is the basis of modern motor-hull design; it conduces to speed in the upright position and neutralizes the strong tendency to draw down at the stern when a motor yacht is driven fast. It can also be adapted to allow the hull to plane at high speed. This double-wedge form, the best for power-driven craft, is the very worst for a sailing craft that is always heeled when travelling fast. It is conspicuous in Fig. 28B, the mid-section of a Lowestoft trawler taken from

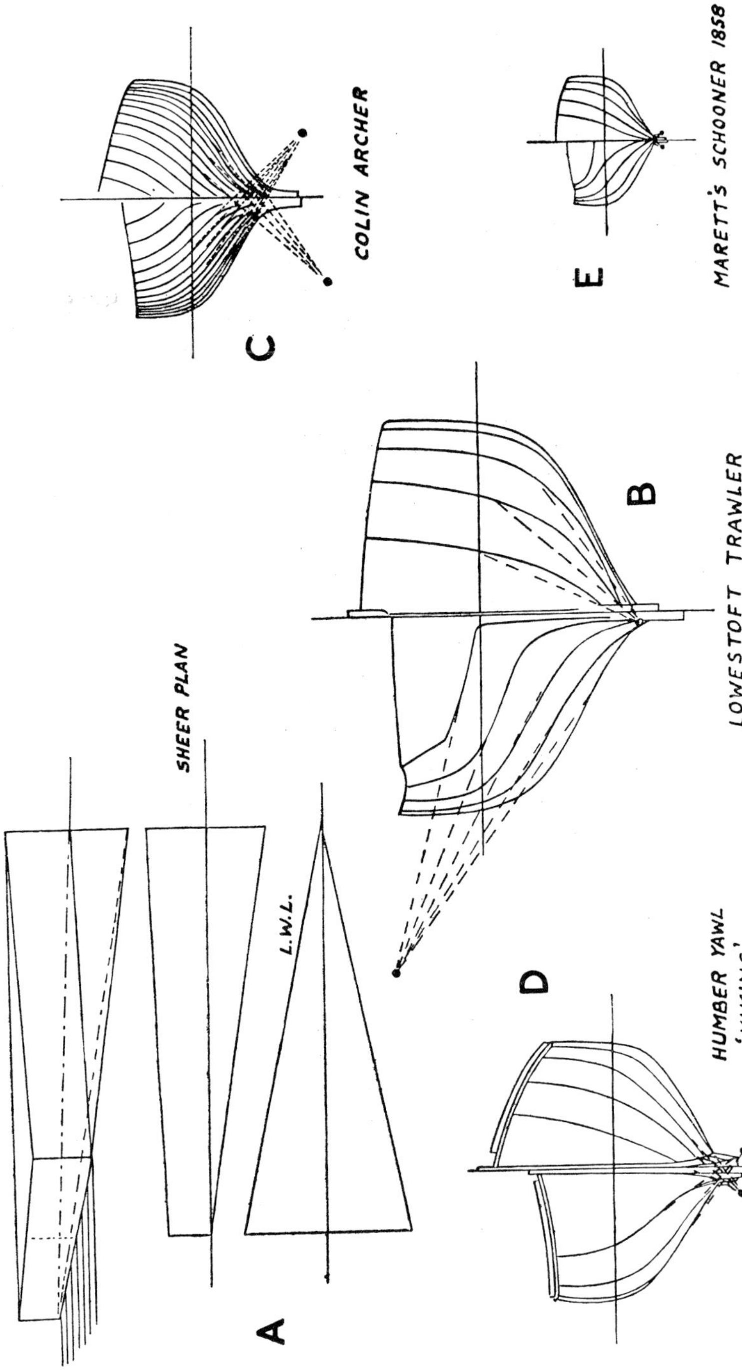

Fig. 28. The double wedge and other hull forms.

Admiral Turner's drawing in his article on the subject in the *Transactions of the Institute of Naval Architects*, 1937, which was copied from a drawing by the late W. M. Blake. It might serve for a Brixham trawler or any other vessel of the type, hatchet bow and malt shovel stern. It will be noted that the bow sections spread fanwise from a point below the water-line, whereas the stern sections radiate from a focus placed laterally above the L.W.L. Compare the trawler with Fig. 28C, which is a Colin Archer type designed by W. H. Rowlands, and with Fig. 28D, the old Humber yawl, *Viking*. These hulls are balanced and their sections radiate from symmetrically disposed foci below the L.W.L. The symmetrical fan is even more obvious in the design of *Sinah*.

Comparing the two forms, the British and the Scandinavian, it is manifest that the bow and stern do not harmonize with each other in the first example, whereas in the second there is complete consonance. The necessity for hull balance was realized by a few some eighty years ago. In 1885 Dixon Kemp published the first edition of his *Yacht Architecture*. I discovered it in the Bodleian Library, and nearly fifty years ago it aroused my interest in yacht design. Later on I picked up in a second-hand bookstall a most interesting little book, *Notes on Yachts*, by Edwin Brett, 1869. He has much to say about the necessity of harmony in design. Fig. 28E is the mid-section of a schooner designed by Marett in 1858; she is perfectly balanced metacentrically and by symmetrical fans. Dixon Kemp insisted that the wedges of immersion and emersion must be harmonized, and that their centres of gravity should be approximated. Unfortunately he did not carry his convictions to their logical conclusion and insist that the centres of upright and heeled buoyancy shall be so close together that they practically coalesce. Had I realized this, my earlier designs would have been much better. I must explain what we mean by the wedges of immersion and emersion, or shortly the "in" and "out" wedges. When a yacht heels she immerses her top-sides to leeward and emerses them to windward. She must take out as much as she puts in otherwise the hull would alter its weight, which is impossible. The immersed wedge taken through the intersection of the L.W.L. and the central body-plan axis, is generally greater in area than the emersed; in Scandinavian types with flaring top-sides it is much greater, whereas in steamer sections with deep bilges the two may be equal. In the first case the hull rises bodily to equalize the displacement, but to some extent the tendency is modified by wind pressure forcing the hull downwards. For our purpose we can safely assume that the hull does not rise, and we may draw our heeled water-lines through the central spot. Figs. 29A and B represent the bow

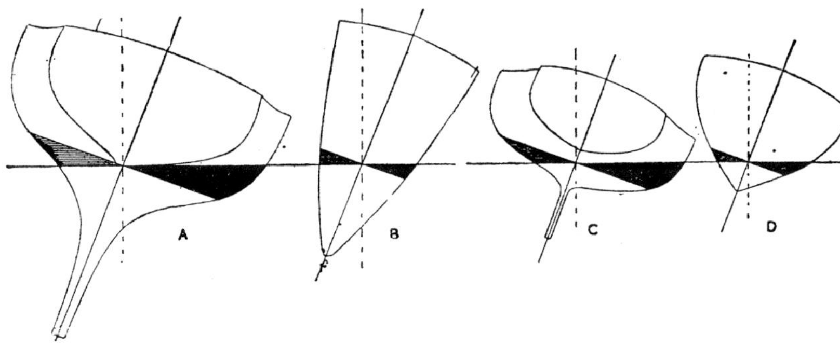

Fig. 29. A and B represent bow and stern sections of a Watson designed fishing yacht; C and D are sections of an X-boat.

and stern of a fishing yacht designed by G. L. Watson. The black area is the immersed wedge; the hatched area the emersed area wedge. In the bow an equal area is immersed and emersed; there is no change that can influence trim. Compare this with the stern section; the hull is putting in more than twice the area that she takes out. It is obvious that the stern must rise to equalize the displacement, and the bow must fall. Every time this hull rolls to starboard or port she dips her head, a pitching movement accompanies the roll, and the motion in a sea-way will be most unpleasant. Compare these sections with those of an X-boat kindly furnished by Messrs. Woodnutt, Fig. 29C and D. Here at both bow and stern the *in* wedge is larger than the *out*, but the proportion is the same. The yacht will rise when she heels, but there will be no alteration in the longitudinal trim. The alteration of trim caused by maladjusted wedges, making a boat root by the head, is one of the causes of griping. I once owned a yacht of the X-boat type designed by Westmacott. She was finger-light on her helm on all points of sailing, even when heeled to the rail. We do not today interest ourselves in the actual wedges because the information they afford can be ascertained in a far simpler manner by the study of the curves of heeled and upright areas and you have already learned how to obtain their centres.

It is obvious from what I have said that to obtain hull balance we must avoid any tendency to the double-wedge form; the sections fore and aft must either be concentric, or they must radiate from symmetrically placed foci. These considerations do not cover the whole ground, for they fail to take cognisance of the profile of the yacht, and this has an outstanding influence upon hull balance. Turner's method of metacentric balance is far more delicate and is a *comprehensive* system for estimating the balance of a hull. He imagines the hull resting upon a plane at right-angles to the inclined water-line, so picture a small model resting upon the edge of a postcard. This card may be a plane superficies as Euclid has it, or it may curve symmetrically one way or the other. Turner calls this plane the *shelf*, because he thinks of the hull as leaning upon it. If the shelf is straight or winds symmetrically on both sides of its mean axis, the model will rest easily and restfully upon it. If, however, the shelf is asymmetrical, crossing the axis x-wise, the hull lies uneasily upon it, wriggles, and tends to slip to one side or the other. This phantasy may or may not help the reader to visualize the conception of metacentric balance. We may consider another analogy, which conveys more meaning to me. The asymmetric hull may be likened to a cone, its base representing the stern of a yacht. If we roll the cone the apex does not move, but the base has a movement of translation; if the cone be rolled to starboard the base will progress to the right, and the bow will tend to point more to the left; in a word the bow is tending to come up to windward. This is in effect the essence of the griping that the double-wedge shows when heeled, the stern swinging away to leeward. If on the other hand the hull is symmetrical it can be compared with a cylinder which when rolled advances evenly with no tendency to swerve from the direction of motion. Whatever the explanation of Turner's system may be, and I confess that I do not fully understand it, I have no doubt that hulls with a symmetrical shelf run straight, they "hold the road," whereas in general those with a crossed shelf do not. I purposely avoid dogmatism for there may be yachts with an indifferent metacentric analysis that are perfectly balanced, and I know of one that is said by her designer to be carefully balanced by every method that we know of, and yet I hear that she pulls hard on her helm. One of my designs has a poor but not a crossed shelf, yet she appeared when I sailed her to be perfectly balanced. Another with practically the same lines, but a larger vessel, pulls hard when reaching fast.

So far we have studied balance from the static side. Turner is emphatic that it is not a dynamic question. He points out that such factors as wave formation and eddy effects cannot be important. The forces that they could generate are far too small in proportion to the total hull weight, in fact he says that they are so insignificant as to be negligible. My mathematical knowledge is quite insufficient to grapple with the question, but one factor must be explained, and it is purely a dynamic matter. If a hull is balanced on a wind she ought to be balanced when reaching. This is by no means the case. I know yachts that balance perfectly when heeled down to the rail by a strong wind when going to windward, but on a reach or with the wind on the quarter they gripe horribly. In the second case the yacht may not be heeled over to the same angle, but the speed is greater and wave formation more marked. Can we neglect wave formation as a factor in balance? I find it difficult to believe that the huge bow wave and quarter hollows that one sees at high speed have no influence upon balance. Has a yacht with a good metacentric analysis a more symmetrical wave formation than one with a crossed shelf? Here photography would give the answer. Admiral Turner really admits that there is a dynamic factor in yacht performance, for he tells us that models will not steer well unless the after deadwood is drastically cut away, and he explains some of the anomalies that I have adduced by the fact that at high speed the after deadwood prevents the bilge stream from getting away quickly and so it is thrown up against the lee side of the counter and affects the steering adversely. The old-fashioned idea that good running ability is conferred by a lot of deadwood aft is being seriously challenged by model practice. The after deadwood is cut away and the rudder hung on a skeg behind it. The triangular profile with drag aft will not do for a model, the fin-keel type run far more steadily. In the model world Turner's system is supreme, and his designs have swept the board. The whole system of balance is a scientific problem based upon theory and practice, and ought to be attacked on the same lines employed in any other research in physics. Unfortunately in the case of real yachts we are faced with two difficulties: only occasionally can the designer discover the real qualities of the yacht he has designed, and in some cases the owner has not an analytical mind, and is easily satisfied, while in others he has the same loyal feelings for his yacht that he should have for his wife. If there are faults he will not admit them till he has sold her—the yacht, not the wife. The second is a far more serious one; in yachting as in medicine, and probably in all spheres of human activity, any innovation, even though it is known to the better informed to be a real advance, is obscured by a fog of prejudice and ignorant opposition which may take years to overcome. This was the case when Lister discovered aseptic surgery. His researches which have brought life and health to countless sufferers were opposed by the majority of contemporary surgeons; they would have had none of it. The metacentric shelf theory is either wrong or right, or partly wrong and partly right. It is not a matter of theorizing, but of actual fact. Do hulls with a crossed shelf or indifferent analysis pull hard on their helms? Are those with a correct analysis easy on the helm? This is the whole question. My own personal knowledge tells me that my older designs, some with a definitely crossed shelf, and others with the "a" curve larger than the "c" curve, do pull hard when reaching rail down. On the other hand my later designs with a correct metacentric shelf are almost self-steering. My experience with *Mystery* designed by Robert Clark, and those described by Mr. Ellis regarding his yacht *Catania*, all point the same way. Not only does metacentric balance prevent griping, but it tends to good running ability. One of the Z 4-tonners which I designed in 1937, 19 ft. on the L.W.L., ran from Beachy Head to Dover before a seventy mile an hour gale in June. Her owner told me that she caused him no anxiety and that she ran beautifully, only shipping one big spray. He was an old square rig skipper, and he said that only off the Horn had he ever seen such a wicked sea. From what has been said it is obvious that we ought to choose a stern that will naturally match the bow. The old-fashioned straight stem can be balanced only with a pointed skiff type stern with an upright stern post. A counter will harmonize with a long spoon bow and a fleet mid-section; it will not mate with the ordinary bow with a short overhang without considerable distortion of the lines. The canoe stern will suit any bow and taken on the whole is the most satisfactory ending for a yacht. The Scandinavian stern is equally sound and has many advantages: it is cheap, easy to build, and eminently sea-kindly. An example is seen in *Thuella* (page 145). The square stern is more accommodating than the counter. The ordinary yacht bow calls for a narrow transom with V-sections to balance the bow sections, one of the Thames barge type, and considerable skill is called for to make this kind of transom look well. It is advisable to draw a pseudo-perspective view of the stern of the yacht; this will give a good idea of what the actual transom will look like. A curved transom gives character to the designs, but except in the very small size it adds considerably to the cost, but in any case a slight curvature is advisable, for if the transom is perfectly flat it has a hollow appearance.

One must not imagine that because a yacht has a good metacentric analysis she must of necessity be a good vessel, for the theory must be used with knowledge and discretion. A rolling-pin and a square box have perfect metacentric shelves, but they are not to be recommended as useful hull forms. A yacht with a crossed shelf could be rectified by affixing a block of wood to the lower part of the deadwood aft, but one cannot think that this procedure would improve the sailing balance. The presence and position of a thin steel centre-board cannot have any effect upon the metacentric analysis, but one must imagine that its position has some influence upon the sailing balance of the ship. It is of course, possible that if a hull is properly balanced there may be a large tolerance to the position of the centre-board. I do not know, but if the position makes very little difference the proof of the theory is almost established. The metacentric theory ought not to be used to determine the nature of a design, but only to analyse it when it has been made, so that any obvious faults can be corrected. It must be used with common sense and knowledge and always with the aid of the balance of the two centres of areas. If the metacentric analysis is perfect and if the centres of upright and heeled areas coincide, then we have done all that is theoretically possible to give good sailing balance.

CHAPTER XI

The Metacentric Analysis

In the preceding chapter I have discussed the principles that govern the sailing balance of a hull; those that tend to make it run straight, to "hold the road," and those that have the opposite effect.

I have suggested that Turner's theory is our best guide to good balance, at least so far as our present knowledge goes, and I now proceed to describe as simply as possible the methods that I employ to plot the *shelf* and from the shelf and the curve of heeled areas to make the final analysis, the so-called *curve of moments*. I originally called this curve the symmetry curve, a term that is entirely non-committal, but perhaps we had better employ Turner's own terminology.

Before describing the actual technique I must first enunciate the symbols that are used: M = Upright transverse metacentre; M.H. = Heeled metacentre; M.A. = Metacentric axis; H.M.A. = Heeled metacentric axis; B = Upright centre of buoyancy; H.B. = Heeled centre of buoyancy; O = Ordinate; L.W.L. = Load water-line; W.L. = Water-line; M.B. = Distance from M to B, this is not the *metacentric height*, both Dixon Kemp and Skene are quite definite on this point, in fact Kemp mentions specifically that M.G. and not M.B. is the metacentric height. On the other hand Symonds in his book *An Introduction to Yacht Design* says that M.B. is the metacentric height, and I have a feeling that I myself have held the same erroneous view.

We have already seen (Fig. 3, page 23) that when a hull heels the centre of buoyancy moves out to leeward of its original position which was on the central axis below the L.W.L. The upright metacentre is found by inclining the hull through a small angle, not more than five degrees, and raising a perpendicular from H.B. to the central axis of the cross-section of the hull. The spot where the two lines meet is the metacentre (M). For such a small area graphic methods to find M.H. would be inaccurate, so they must be calculated. At larger angles of heel the M.H. may not be in the same position as M. If M.H. is vertically above or below M, or if the two spots coincide, then Turner calls the hull a metacentroid. A hull may be a metacentroid and yet be quite unbalanced.

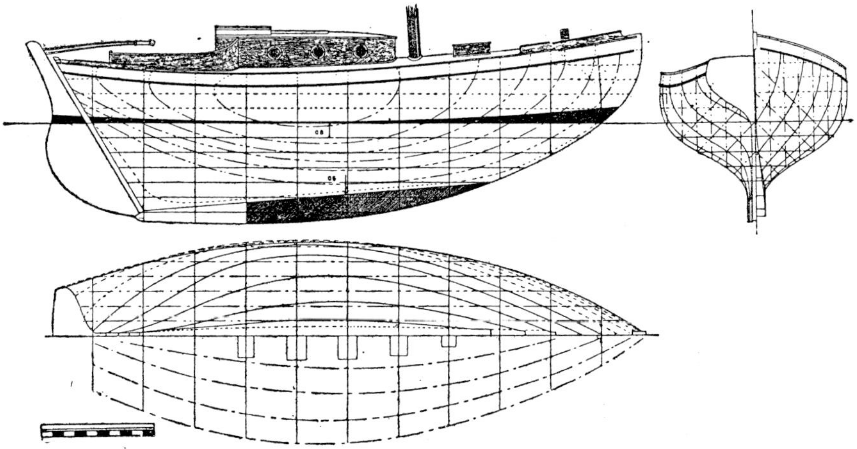

Fig. 30. *ENGLYN*, **a 7-ton cruiser.**

I propose to use my design *Englyn* to demonstrate the analysis. She is a 7-ton yacht having the dimensions shown in Fig. 31, and six or seven or even more have been built in this and other countries. The same analysis applies to the design *Askadil*, because the two yachts have identical sections. I have sailed in two of the *Englyns* and in *Askadil*, and I know that although their analysis is not perfect, they are in fact reasonably well balanced, and can be made to sail to windward unattended.

The drawing of *Englyn* (Fig. 30) shows that the L.W.L. is divided into ten equal parts making the spacing of the sections 2.25 ft. Her water-lines are spaced 8 in., or 6 in. if we measure them with a scale of an inch to a foot. The scale of the design of *Englyn* is $\frac{3}{4}$ in. = 1 ft.

Our first procedure is to find the position of B below the L.W.L. To do this we have to find the area; first of the L.W.L. plane and then of each successive water-line plane, from L.W.L. downwards. In Fig. 31B I have labelled them, L.W.L., and then A, B, C, D, etc. The measurement of these areas will be most laborious without a planimeter, for each will have to be calculated either by Simpson's or the Trapezoidal rule. The former is the more accurate, but the latter is simpler and sufficiently exact for our purpose. I have already explained the Trapezoidal rule, but it will save the reader trouble if I recapitulate the procedure. To find the area of a plane superficies rule a base-line across it; in this case it is the central line of the half-breadth plan of the hull, and divide the line into a convenient number of equally spaced parts, and at each station raise a line normal to the base-line. Now using the scale to which the yacht is designed, in this case $\frac{3}{4}$ in. to 1 ft., measure the length of each of these ordinates, add them up, subtract half of each end ordinate, and multiply the remainder by the spacing of the ordinates. The product is the area required. In the case of the water-line planes under consideration the

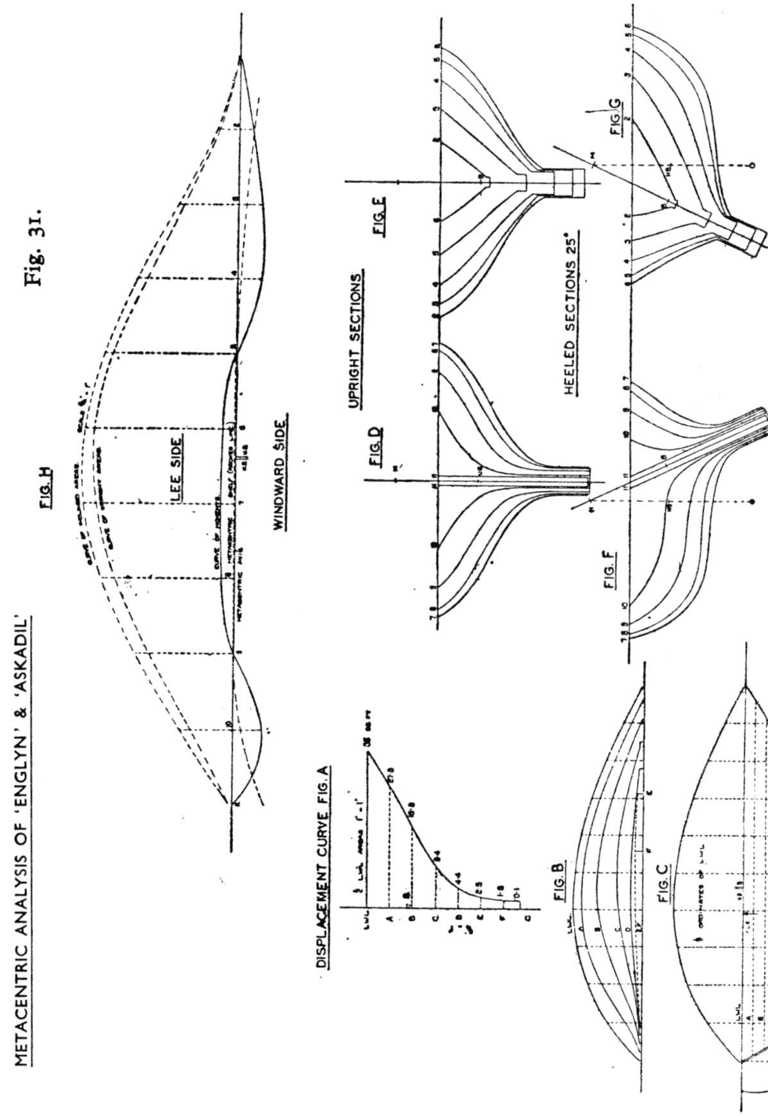

Fig. 31.

curves end in a point, so all that we have to do is to add the ordinates and multiply by the spacing. The L.W.L. is already divided into ten parts, and we can use the divisions on the drawings as ordinates, but the rest of the planes will have to be divided up by suitable ordinates. But as I have already said a planimeter is almost essential in yacht designing. My own planimeter belonged to the late Albert Strange and was used in the design of *Tally Ho*, the winner of a Fastnet race, and for other notable craft. It has no adjustment for scale and reads only for a drawing of one inch to a foot. If the drawing has another scale we multiply by a factor. In the case of Englyn, whose scale is $\frac{3}{4}$ in. to 1 ft., we multiply by 4/3 squared, that is by 16/9. If the reading indicates a volume, multiply by the inverted scale cubed.

For our present purpose no corrections for scale are required, for we are dealing with proportions and not with absolute dimensions, so we can write down the figure as it is read off from the planimeter. If the instrument is one of the small variety do not stretch it out to its full extent. Place the planimeter upon the original drawing and measure the area of each water-line plane as far as the mid-section. Write down the figures in a row. Now repeat the process for the other half of the water-line plane, and place the figures under the first set. Thus we have the areas of two half water-line planes under each other down to the keel. Add the two sets of figures together and we now have the half area of each water-line plane. In larger drawings we divide the planes into three sections and add the areas indicated. We have now to make a vertical displacement curve as in Fig. 31A. Rule a vertical line as an axis and a normal to it, a horizontal top ordinate, in this case the L.W.L. Then mark off from the scale to which the yacht is designed, in this case $\frac{3}{4}$ in. to 1 ft., a series of ordinates spaced to the same distance as the water-lines of the sheer-plan of the yacht. In this case they are 8 in. Along each ordinate mark off from the scale 1 in. = 1 ft. the areas of the water-line planes that we obtained with the planimeter. In the case of *Englyn* they are: 35, 27.3, 18.2, 9.4, 4.4, 2.5 and 1.8. Join the markings with a fair curve, which is the *vertical displacement curve*. Now the water-lines on the sheer-plan do not include the lowest part of the keel aft. So continue the displacement curve downwards and end it with an ordinate which is placed at the mean draught of the keel below water-line F. This is water-line G, and I have estimated the area as 0.1. This vertical displacement curve is a very important one. It will be found in the design of *Sinah* and of all the designs found in this book, in fact no design can be considered complete without it.

Not only do we obtain from it the centre of buoyancy considered in a vertical plane, but a simple calculation gives the total displacement of the yacht, and we can easily find out from it the effect of adding or removing weight upon the draught of the hull. The calculation for displacement is given with the other calculations at the end of this chapter.

To find the centre of buoyancy, take a piece of good tracing or detail paper and on it trace the curve of displacement, including the ordinates A and B. Then poise the curve on a razor edge and it will balance on the vertical centre of buoyancy. Make a mark and rule a pecked line through the mark parallel to the ordinates. Now poise again to confirm the balance and to make certain that it is poised along a line parallel to the ordinates. Before poising it will be necessary to crinkle the curve. Place the edge of a scale at right-angles to the ordinates and fold the curve over the edge of the scale making a tight crinkle. A second crinkle may be necessary. These, like the corrugations in a sheet of galvanized iron, will prevent the paper from bending. Take the greatest care in poising. A mere breath of wind will upset the balance. If the designer thinks that the position of B ought to be calculated he can use the method we have already described, taking moments by Simpson's rule or by the amended Trapezoidal rule. As in the case of keels, taking moments in the ordinary way with the Trapezoidal rule will introduce a definite error. The line of balance runs through B on the axis. We have now discovered the location of the centre of buoyancy in the vertical plane. Transfer this position from the displacement curve to the body-plan and the sheer-plan. A pair of tweezers of the kind that women use to mutilate their eyebrows will be found useful to hold the thin bits of paper during poising.

We have now found our first fundamental spot, and whereas the whole analysis is founded upon the centre of buoyancy, it is impossible to be too careful, by taking the mean of more than one planimeter reading, in cutting out the graph, and in poising it.

The next stage is to find the metacentre. This involves a very simple calculation. The rule is as follows: *The height of M above B is equal to the moment of inertia of the L.W.L. plane divided by the volume of the displacement.* This sentence sounds very learned, but it's mere jargon to me and probably to many of my readers. But reduced to ordinary language it is quite simple. Take half the widths of the L.W.L. plane; cube them, and add the cubes together. Now divide their sum by three times the sum of the upright half-sections. The reader will remember that the upright half-sections have already been used to calculate the displacement of *Paida*. At the end of the chapter they will give us the displacement of *Englyn*, and this can be compared with the displacement obtained from the areas of the water-line planes. The two should of course be the same.

Here is the formula in equation form:

$$\frac{\text{sum of cubes of half L.W.L. widths}}{3 \times \text{sum of half areas of upright sections}}$$

My readers may ask: why divide by 3? The only answer that I can give is just "mathematical joss." The figures for *Englyn* are:

$$\frac{133.322}{3 \times 24.27} = 1.83$$

That is to say the metacentre of *Englyn* is 1.83 ft. above the centre of buoyancy, using a scale of 1 in. = 1 ft. The actual distance B.M. is:

$$\frac{1.83 \times 4}{3} = 2.44$$

because the scale of the drawing is $\frac{3}{4}$ in. = 1 ft.

The cubes are obtained from a table of cubes. It is quite easy to go wrong with the table, for it gives no *place-values*. If we begin in the middle of the column of the half water-line ordinates and work up and down, we shall at once see whether we have placed the cubes under each other in the right position. We need not trouble about the decimal point, for we know the answer within a little.* If, for example, B.M. came out at 18.3 it would obviously be wrong, whereas 1.83 is seen to be correct. I hope that my mathematical readers will excuse me for trying to make the matter quite plain, but when I began I found it took me quite a little time to learn how to use the cube tables. Of course it can all be done with a slide rule or a table of logarithms, but the cube table is quicker. Now take a scale of 1 in. = 1 ft. and measure 1.83 in. up from B and we have the metacentre. Note that in all these calculations we have assumed that the scale is 1 in. = 1 ft. because our planimeter has no corrections of its own. Therefore we must measure the height of M with a scale of 1 in. = 1 ft. Some more expensive planimeters have a correcting screw by which the readings can be made on varied scales. Later on we shall have a check upon the position of the metacentre, for we shall obtain a prometacentre, and if this is far from M then the calculations for M are wrong.

We have now discovered the position of the transverse metacentre, the central spot for our analysis. There is, of course, a longitudinal

* The position of the decimal point is readily seen by inspection. The points have, for the sake of clarity, been inserted in the table on page 118.

metacentre, but this does not concern us. M, of course, lies along a straight line that runs along the yacht from stem to stern in its axis plane.

We have now to locate the shelf. In an earlier chapter we made a rough location of the shelf using the Welch axis; now we shall elaborate the method using the metacentric axis as our base line. To facilitate the next stage it is convenient if the yacht has been designed with two body-plans, one bow port and starboard and one stern, also port and starboard. I have (in Figs. 31D and E) done this, for that part of the body-plan that is under water. We have already seen how to draw the inclined water-lines when we plotted the curve of heeled areas. Be very careful that the two lines are inclined at the same angle. It is not sufficient to trust to a protractor. When you have ruled in the water-lines test them by measurements on two equidistant buttock lines. In Figs. 31F and G I have drawn the inclined water-lines of *Englyn*. We incline the yacht to the rail, or near to this inclination because if the analysis is good at this extreme angle it will *a fortiori* be even better for smaller angles of heel. It is obvious that when a yacht of the orthodox shape heels she will rise bodily to maintain the same displacement, and that it is therefore theoretically incorrect to make our new water-line cross the central axis at its original location. But we may assume that the wind force that heels the yacht will also depress her bodily and to some extent neutralize the tendency to rise. I think that it is sufficiently accurate to let the inclined water-lines cross at the spot where the upright water-lines cross the central axis. Now from M drop a perpendicular to the inclined water-line and continue it downwards across all the sections. This is the metacentric axis, a section of an inclined plane that passes from stem to stern, the metacentric plane. We have now to continue our analysis to discover to what extent the volume of the hull is disposed symmetrically with regard to this plane.

Now take the body-plan, preferably the double port and starboard plan, and trace on good tracing or detail paper all the forward sections as shown in Fig. 31G. Rule in the metacentric axis and the central axis. If we are dealing with the ordinary single body-plan we first trace the lee sections and then turn the paper over and adjust it to the central axis, the inclined water-line, and the metacentric axis, and trace the windward sections. The lee and windward sections are traced on opposite sides of the paper, but this does not matter for the paper is transparent. Carefully cut out the outside section and poise it on the razor. Mark the line of poise. Rule a convenient base line on the main drawing parallel to the water-line and drop down the ten ordinates from the sheer-plan. It is wise to number all these stations and those of the body-plan. It will be convenient to place this base line, which is the metacentric axis, close to the curve of heeled sections for a reason that will shortly appear. Now mark off the distance of the line of poise from the M.A. and transfer it to the base line that we have just drawn. This can be done by approximating the traced section to the new base line. Now cut off the section that has been poised and poise the new section, and so on till the whole of the bow sections have been poised and the distances transferred to the base line. It is quite unnecessary to do as I used to do, make a fresh tracing for each bow section. One tracing will do for the whole lot because each section diminishes all round. The same process will have to be done for the stern, but generally we shall have to make a separate tracing for each section. This is because the sections diminish, but the heels of the sections lengthen. When all the sections have been poised and the poise distance transferred to the base line as in Fig. 31H we have a series of spots. Run a curve through these spots and this is the shelf. If there is any difficulty in obtaining a shelf with a reasonable degree of harmony, it will be well to retrace any section that appears to be out of the true curve and repoise. If the yacht is itself symmetrical the shelf ought to be equally symmetrical, but in the case of yachts with an unsymmetrical profile like *Paida*, the shelf will not be so sweet as in those with a profile with an unbroken curve like the *Khamseen A*. In Fig. 31H the windward curve has been drawn below and the lee curve above. We now have a shelf that is exactly the same as that given by the Welch method.

The final stage is to make the curve of moments. In the article that appeared in the *Yachting Monthly*, Vol. LXII, page 200, I described a method for finding the prometacentric shelf by gumming all the sections together and poising them *en masse*. This is Blom's method, but both Turner and myself have found it to be inaccurate. We now adopt the old-fashioned method of trial and error.

If the original drawing has been made on ordinary paper, plot out the shelf against a metacentric axis ruled on tracing or detail paper, and use thin lines for axis and shelf. It is convenient to use red ink for the shelf. Lay the shelf alongside the curve of heeled areas and with a scale that has small divisions, such as ⅛ in. to 1 ft. or a centimetre scale, measure the length of ordinate 1 of the heeled areas and the distance between the shelf and the axis. This distance is called the *discrepancy*; multiply the two together and transfer the product to the same ordinate on the axis, using the same scale. It is most convenient to have the two drawings, that of the curve of heeled areas and that of the shelf, side by side as in Fig. 31H, because both measurements, that of the discrepancy and the area of the heeled section that corresponds, can be taken *seriatim*, and there is no likelihood of any muddle. This can easily happen if the areas are read from a column. Before we start plotting the shelf we should print "Lee" on the upper side of the axis, and "Windward" on the lower side. As we poise the sections we shall find that the line of poise is either to windward or leeward. If the design is of the ordinary cruising type the forward and after discrepancy will be to windward, the central discrepancies will be to leeward. That is to say that the shelf will wind to leeward amidships and to windward at the bow and stern. If the yacht is of the racing type or an ocean cruiser with deep keel amidships, then the positions will be reversed. We shall revert to this point later. In a few rare cases the shelf will lie more or less parallel to the axis. It may coincide with it if the hull is a metacentroid, or it may lie to leeward or windward. When we poise the sections the line of poise will be outside the metacentric axis in the fore and aft sections and inside amidships. This, of course, is the cruising type. When we have completed all the multiplications and have plotted the products to windward and leeward of the axis we have a series of dots. We now join these dots by a curved line which winds from windward at the fore end to cross the axis and again returns to windward. If this curve is a very unsymmetrical one we must overhaul the measurements and the multiplication and try to get it as symmetrical as possible. We must realize that if the hull is symmetrical, the moment curve must also be symmetrical, apart from certain reasonable irregularities due to an asymmetrical profile. These three curves form the *curve of moments*. We call the forward one $-a$, the amidship curve $+b$, and the after curve $-c$. This in the cruiser type of curve.* In the case of raters and similar craft, including the *Victory*, we label the cruves: $+a$, $-b$ and $+c$, or simply a, b and c. If the diagram shows a crossed shelf there will be no b; simply $+$ or $-a$ and $+$ or $-c$.

It is a law of hydrostatics that the area of $a+c$ must equal b. To obtain proper balance, a must be equal to c. If a is not approximately equal to c, then the lines of the hull must be modified. I shall show how this can be done in Chapter XIII. If $a+c$ is not equal to b then we must rule another axis very close to and parallel to the original metacentric axis, and begin the measurements and multiplications all over again. If the central curve is too small, then the new axis will be made on the side of the M.A. away

* See moment curve on page 50 *ante*. It will be seen that discrepancies to windward are considered minus and *vice versa*.

from the concavity of the curve to increase the discrepancies, and *vice versa*, but this is a matter of common sense. It may be thought that the process is very tedious, but with practice one or two trials will hit the mark. A slide-rule will accelerate the calculations. As a matter of fact an absolute satisfaction of the equation is unnecessary. How far we can depart from the equality of a and c is a matter which must be decided by time and experience, but I think that a must never be larger than c or the yacht will be hard-headed. Probably c may be considerably larger than a without causing lee helm. We have now located the final axis that makes $a+c$ equal to b. This is the prometacentric axis of the H.M.A. If we carefully measure the distance between the new H.M.A. and the old M.A. with accurate dividers, we can rule in the H.M.A. parallel to the M.A. on the body-plan, Fig. 31F and G. If the two do not coincide the hull is not a metacentroid.

The discrepancies amidships are very small and they have to be multiplied by the large areas found in this part of the hull. It is therefore obvious that if the very greatest care is not taken, large errors can appear. For this reason the graph must be made of paper with a smooth surface. I use the best detail paper that I can buy; before the war it was a Belgian make. The lines must be ruled with a fine ink line and those that are tentative with an HHH pencil. I use a lens to read off the measurements on the scale and on the planimeter. The scale must be a good one, in fact it pays in this sort of work to buy the best instruments. Rieffler's are the best, but it may be a long time before they can again be obtained. The same applies to paper. When judging competitions I have been amazed at the cheap and nasty paper used by some of the competitors. An extra shilling or two is well spent upon work that may take some weeks to complete. When we have done our best the analysis is not strictly accurate, but if we are careful it is good enough for our purpose. Fortunately the inaccuracies are most likely to appear amidships where they are not so important. The chief object is to obtain curves a and c so that we may be able to compare their areas. We want to obtain a curve in which a is approximately equal to c, and to be careful that c is not less than a.

THE COMPLETE CALCULATIONS FOR THE ANALYSIS OF *ENGLYN*

CALCULATION FOR METACENTRE

	Areas of Half Upright Sections	*Half Widths of L.W.L.*	*Cubes of Half L.W.L. Widths*	*Areas of Heeled Sections*
1	0.00	0.000	0.000	0.00
2	0.46	0.850	0.614	1.16
3	1.27	1.660	4.574	3.50
4	1.55	2.300	12.170	5.90
5	2.65	2.700	19.680	7.89
6	3.66	3.000	27.000	9.14
7	4.22	3.050	28.370	9.20
8	4.23	2.855	23.370	8.14
9	3.60	2.400	13.820	6.00
10	2.63	1.550	3.724	3.80
11	0.00	0.000	0.000	0.00
	24.27		133.322	

$$\frac{133.322}{24.27 \times 3} = 1.83 = \text{BM (Height of Metacentre over Centre of Buoyancy)}$$

$$\text{Correcting for Scale, BM} = 1.83 \times \frac{4}{5} = 2.44 \text{ feet}$$

CALCULATIONS FOR DISPLACEMENT

A. BY TRAPEZOIDAL RULE, FROM L.W.L. AREAS

Half L.W.L. Areas	
L.W.L. ..	35.0
A	27.3
B	18.2
C	9.4
D	4.4
E	2.5
F	1.8
G	0.1
	98.7

$$\text{Displ.} = (98.7 - 17.5) \times \frac{2}{3} \times \frac{16}{9} \times \frac{2}{35} = 5.5 \text{ tons}$$

(Areas in square feet. Ordinates in feet)
(Scale: 1 inch = 1 foot)

98.7 = Sum of ½ L.W.L. areas × 2 = sum of areas
17.5 = ½ sum of top and bottom figures
$\frac{2}{3}$ = Spacing of areas in feet
$\frac{16}{9}$ = Correction for scales
35 cubic feet of sea water = 1 ton

B. FROM AREAS OF UPRIGHT SECTIONS

$$\text{Displ.} = 24.27 \times 2 \times \frac{16}{9} \times \frac{2.25}{35} = 5.55 \text{ tons}$$

24.27 × 2 = Sum of areas of sections
$\frac{16}{9}$ = Correction to scale of ¾ inch to 1 foot
2.25 = Spacing of sections
35 cubic feet of sea water = 1 ton

CHAPTER XII

Finding the Centre of Gravity

By Lieut.-Cdr. A. L. Braithwaite, R.N.V.R., A.I.N.A.

I use the following method for finding the weights and centre of gravity of a yacht, and from that determining the weight and position of the centre of gravity of the lead keel, so that the yacht shall trim as designed.

This is a matter of great importance in the type of yacht I was designing and building for ten years before the war. These are light displacement craft in use on the Broads, with a fin keel and all the ballast on the keel, at least that is where it should be. In my first year I was surprised at the number of boats I saw with varied weights of inside ballast. This is not only a great nuisance when boats come back with milk or paraffin spilt in the bilges, but it is not exerting anything like the effect it should if it were in its proper place on the keel. If a yacht is properly designed there is no excuse for this; by properly designed I mean on the drawing-board, with displacement and centre of buoyancy worked out before building starts. The best way of building these boats is on a stock on which they remain till practically completed, so that the keel is the final stage of the work; from this it will be obvious that one need not worry about the shape and weight of the lead keel till its weight and the position of its centre of gravity can be definitely determined. This is a great advantage, as it is certain before the boat is launched that she will float and trim as designed, and one does not have the extremely annoying mishap of finding that the line of the boot topping is either submerged or too high. The line is usually scribed while she is still on the stock, when a good line is easily marked, but it is far from easy to re-mark it properly when the boat is afloat, in fact it is practically impossible, and if it is necessary, it is a very definite sign of bad designing. The process I am about to describe is perfectly simple; all that is necessary is a spring balance, weighing up to 200 lb., hung in the building shop, and the co-operation of the builder who must be persuaded religiously to weigh each part as it is ready to be built in. If the reader is himself the builder, there is no difficulty, but if a professional is doing the work he will doubtless say that his great-grandfather didn't do it like that, and he may regard you as a mild form of lunatic, but if you want a good result, insist that it must be done. I am quite sure that my own men thought I was a little mad, but I was always in the building shop during building operations, and they soon saw that there was method in my madness. It is worth while to explain to them why you want it done; you are much more likely to gain their co-operation if you do. A sheet should be pinned up in the shop close to the spring balance, and as each part is ready it is weighed and entered on the sheet, which, when full, is transferred to the drawing office. It is of course unnecessary to weigh the planks on both sides, each plank is weighed on one side only, and similarly all parts such as stringers, etc., which are alike on both sides, but it must not be forgotten to double these weights. It will be found that very soon this routine becomes automatic and the delay caused is negligible. It might be thought that these weights could be worked out from a table giving the weight of a cubic foot of whatever timber is being used; so it could, if anyone is possessed of superhuman perseverence and patience, but the result would not be nearly so accurate and cannot be considered a practical method.

When all the weights are known in the actual boat, it is also necessary to get the weights of spars, sails and crew. It is presumed that the spars will have been made while the building has been proceeding, and if not too heavy the weights can be got from the spring balance, but failing this the mast weight, which is the only one that may be too heavy, can be got from the following table:

WEIGHTS PER FOOT OF LENGTH OF SPRUCE SPARS

Diameter	*Weight*	*Diameter*	*Weight*
3 in.	1.62 lb.	$4\frac{1}{4}$ in.	3.25 lb.
$3\frac{1}{4}$ in.	1.90 lb.	$4\frac{1}{2}$ in.	3.65 lb.
$3\frac{1}{2}$ in.	2.20 lb.	$4\frac{3}{4}$ in.	4.07 lb.
$3\frac{3}{4}$ in.	2.53 lb.	5 in.	4.50 lb.
4 in.	2.88 lb.		

The table has not been carried beyond a five-inch mast, as it is unlikely that any yacht built by an amateur will require a heavier mast; incidentally it is worth noting that for a given strength a hollow mast will be slightly more in diameter, and roughly half the weight. If we are dealing with a mast which can be lowered, it will require a balance weight, and this is found by placing the mast on two trestles with a triangular piece of wood under the position of the mast pin. The mast rests on this, and slabs of lead borrowed from another mast are piled on till the mast head just starts to lift; any weights of course will do, but they must be placed on the heel of the mast or a false weight will be obtained. The spars can be weighed on the spring balance, also the sails, and the crew are averaged at ten stone. We now have all that is required, and proceed to rule up a table as shown later. On this we fill in the names of all the parts, and their weights taken from the rough sheets entered up in the shop.

A short spell of rather laborious but perfectly simple work will now have to be endured, and we shall be nearing the object we have been aiming at, viz. to find out the weight and position of the centre of gravity of our lead keel. It is most important that the crew should be included if the boat is of the skimming dish type, like *Kala-Nag* described in *The Yachtsman* and built in 1939. She was almost as sensitive as a dinghy to the movements of her crew, and when they were on board her trim altered materially, so that in multiplying their weight by the arm (which is the distance from the centre of buoyancy measured on the plan) it was necessary to determine where they would sit. If we are considering a 30-ft. cabin cruiser the position of the crew is not nearly so important, and their weight multiplied by the distance of the centre of buoyancy to the centre of the well is quite satisfactory. To continue then after this slight digression, each item in the boat's make-up must be taken, and its position in the boat from the centre of buoyancy measured in feet and decimals of a foot. Then multiply the weight of each by the arm, and enter the result under moments forward of C.B. or aft of C.B., according to whether the part is forward or abaft the C.B. as the case may be. In the case of the planking, stringers, or in fact any part which extends equally both sides of the C.B. the weight only is entered, as there is no moment, the weight being equally distributed both sides of the C.B. The rudder in this type of boat is specially important as a balanced rudder is fitted; this, of course, is not attached to the keel at all and is far aft. As these are of considerable size and made of boiler plate, they frequently weigh half a hundredweight or more and may be 12 ft. or more from the C.B.; it will be appreciated that they influence the trim very definitely.

A full water tank of probably 10 or 12 gallons is often located in the stern locker, and the crew in the well all help to put the boat very much down by the stern, but against this we have the mast and its balance weight (Broads' yachts do not carry an anchor and cable). Altogether the C.G. of the entire yacht is generally found to be from 6 to 9 in. abaft the C.B. and to find it we add up the column of weights, and those of the moments fore and aft, subtract the lesser total of the moments from the greater, and divide the result by the sum of the weights. This gives the centre of gravity, and it is forward or abaft the C.B. according to which

LIST OF WEIGHTS AND MOMENTS

C.B. 1 INCH FORWARD OF STATION 6

Item	*Weight*	*Arm.*	*Frd. of C.B.*	*Aft of C.B.*
Keelson	60	—	—	—
Stem	25	11.5	287	—
Transom	4	12.25	—	49
Stern Knee	2	12	—	24
Planking	340	—	—	—
Floors: 1	7	6.75	47.25	—
2	7	1.75	12.25	—
3	7	—	—	—
4	8	2.5	—	20
5	7	6.75	—	47.25
Stringers	48	—	—	—
Mast Beam	47	4	188	—
Main Sheet Beam ..	9	11	—	99
Aft Well Beam.. ..	9½	9.5	—	90.75
Tabernacle	17½	4.3	75.25	—
Mast Step	22	4.3	94.5	—
Tabernacle Accs. ..	12	4.3	51.5	—
Tie Beams	20	5	100	—
Samson Post	8	9.5	76.0	—
Hatch Runners ..	21	6.5	136.0	—
Hatch Beams	6	8	48.0	—
Half Beams each ..	3	6.5	78.0	—
Breast Hook	2½	12	30.0	—
Lodging Knees ..	10	4.3	43.0	—
Hanging Knees ..	5	4.3	21.5	—
Quarter Knees.. ..	3	12	—	36
Bulkheads	10	2 } 4.5 }	—	20 } 45 }
Fore Peak Floor Boards	50	7	350	—
Deck	188	116 8 } 62 3 } 10 10.5 }	928	186 105
Cabin Sides both ..	40	1	—	40
Cabin Front	12	4	48	—
Cabin Purlins	36	—	—	—
Cabin Top	112	—	—	—
Rubbing Strake ..	30	—	—	—
Mast	190 }			
Mast Lead	}	4.3	817	—
Hatch Coaming ..	14	6.5	91	—
Rudder complete ..	56	10.5	—	588
Hatch	23	6.25	143.75	—
Main Floor Boards ..	70	—	—	—
Stem Band	10	12	120.0	—
Aft Bulkhead	11	9.5	—	104.5
Mainsail and Jib ..	30	1	30.0	—
Boom and Fittings ..	28	3.75	—	105
Gaff	20	4	80.0	—
Seats	20	7	—	140
Carried forward ..	1660.5		3896	1699.5

Item	*Weight*	*Arm*	*Frd. of C.B.*	*Aft. of C.B.*
Brought forward ..	1660.5		3896	1699.5
Anchor	56	9	504	—
Cable	100	8	800	—
Wood Keel	200	·75	—	150
Centre B. Case ..	20	1	20	—
Centre Plate	256	·5	128	—
Crew	300	6.5	—	1950
Aft Well Step	5	9.5	—	47.5
Sundries, Cushions, Stores, etc.	300	—	—	—
Outboard Engine ..	60	6.5	—	390
Water	124	7.5	—	930
TOTALS	3081.5		5348.00	5167.0
			5167.0	

3081.5) 181.000(.058
154.075

26.9250 C.G. .058 ft.
24.6520 frd. of C.B.

2.2730

Lead Bulbs, 560 lb.

$$\text{Arm} = \frac{181}{560} = .32 \text{ ft. abaft C.B.}$$

Weight of Hull, etc. .. 3081.5 lb.
Add 560 lb. Lead .. 560.0 lb.
Total Weight .. 3641.5
Displacement 3808.0

166.5 lb. available for additional crew

moments were the greater. Now we are really getting somewhere; we know the total weight of the yacht without keel; we know the total displacement at the designed W.L. previously worked out, and their difference is of course the weight of keel required, part wood and part lead. The centre of gravity of the lead keel must be in such a position as to balance the amount the C.G. of the whole yacht is forward or abaft the C.B. Perhaps the best way to make this clear is to take the actual calculation made for *Restless*, which was entered in the competition for a 20-ft. L.W.L. centre-board cruiser in *The Yachtsman*. This boat has not been built, but the weights were taken from those obtained during the building of another boat, which had the same dimensions, though different lines, and the figures may therefore be taken as accurate for all practical purposes.

It will be noticed that the C.G. of the hull and gear is forward of the C.B., *not* abaft it as in all the other yachts I have designed. This is probably due to the fact that *Restless* carries an anchor and cable, whereas, as previously stated, none of the Broads' yachts are fitted with one. It will be seen from the table that the combined moments of these is 1304 forward of the C.B., a considerable item. This only serves to emphasize the importance of the process I have described, as had it been assumed that the C.G. was about 6 in. abaft the C.B. as usual, the C.G. of the lead keel would have been required to be forward of the C.B. and the boat would certainly have been badly down by the head. Also both centres of gravity are much nearer the C.B. than I have previously found to be the case, so that guessing will be almost certain to lead one badly astray. Instructions for arriving at the pattern for the lead keel are fully dealt with elsewhere in this book, so that it is unnecessary to say more on this subject, but a description of the method of assembling the keel and its fixing to the hull may be of interest. The wood keel should not be made till the pattern for the lead is satisfactory; it can then be fitted to the pattern to give the profile as designed. Before being sent to the foundry (and if there is one near it will certainly pay the amateur builder to get it cast) the exact position of the keel bolts must be clearly marked, stating what diameter bolts are to be used. The holes are then cast in the keel, and a great deal of time and hard work will be saved, while the extra cost is not worth considering. Do not be misled into thinking that you can bore through lead like a piece of cheese; boring a hole probably 6 or 9 in. long and $\frac{5}{8}$ or $\frac{3}{4}$ in. diameter is not a job to take on if it can be avoided; it is quite enough to bore the holes in the wood keel when the lead comes back from the foundry. The first operation is to remove the boat from the stock and support her at each end, high enough off the ground to allow the keel to be placed underneath, with enough clearance for the bolts; four casks stood on end, two near the bow and two near the stern, are very satisfactory. A piece of stout stuff about 2 in. by 6 in. is laid across each pair, with packing pieces on the top; the boat is then moved sideways on to these; two or three men can lift one end across at a time, and some rough props are put in if necessary, though it will be found that she will remain upright without any difficulty. The top plank of the keel must now be offered up to the hull, being careful to see that it fits well and is exactly in the right position; the bolt holes are then marked through those already in the keelson. Having accomplished this very important matter, the whole keel is assembled on trestles (not forgetting to paint the edges which come together with red lead and varnish), cramped up with sash cramps, and a line drawn where each bolt comes from the centre of the hole in the keel, to the centre of the mark on the top plank. The holes are then bored, working from both top and bottom; if someone with a straight eye can be found to stand some distance away and say if you are going straight it will be a great help.

It is hardly necessary, I suppose, to emphasize that great accuracy is necessary in this, and also in boring the holes in the keelson which is done much earlier; in practice these holes are bored when the keelson is set up, and they are used where possible to bolt it to the stock. As soon as the holes are well and truly bored, the most difficult part of the job is done; the bolts, well greased, are driven in and the keel is ready to go on. By the way, the bolts, which you will have to get a blacksmith to make, will have square heads, and a socket for these will have been cast in the lead, but before driving the bolts, be very careful to check your measurements (they will all be different lengths), and make sure there is enough thread on them to allow the keel to pull up tight. A mistake in this may not otherwise be noticeable till the keel is almost on, and you will be in a real mess, as you can do nothing but take the keel off again to rectify the error. We will assume that you have taken all precautions, and the keel is now stood on edge and ''walked'' under the boat, using a lever to get it in position; pack it up till the longest bolt is almost touching, and exactly under the hole it is to enter. One end of the boat is then lifted enough to get a piece of packing (about an inch thick) out; the boat is let down on the remaining packing, and the bolt should enter; the same performance is then carried out at the other end, it being most important to keep both ends level. Continue lowering till all bolts are entered, and there is only one further precaution to take before settling her hard down on the keel. Each bolt must have a piece of caulking cotton soaked in red lead wound round it, and when the bolts are far enough through a similar piece on the top of the floor under the plate on which the nut pulls down; this is, of course, to guard against the possibility of a leak. The boat can now be worked down, doing a little at each end alternately, and removing packing pieces. It may be that when she is nearly down you will get stuck if the bolts are a tight fit as they should be; the only thing to be done then is for one hand to get inside (be sure to prop her up firmly while he gets in), and while she is held upright use a little persuasion with a large maul, being careful to hit the floors; as soon as the nuts can be well entered they will do the rest.

One final tip, don't putty up the seams till the keel is on, as in view of the aforesaid persuasion it will *probably be loosened, and you will have to go over it all again.*

It should be noted that this chapter deals only with yachts carrying *all* their ballast on the keel. I have built these up to 3 tons, and most amateurs are not likely to try anything larger. If we consider a sea-going yacht with only part of the ballast outside, and enough inside to enable the proper trim to be obtained, it is not nearly so important to find the exact weight of the keel, and its C.G., as the ballast inside can be trimmed as necessary. Indeed if you have a lead keel weighing 2 or 3 tons you can hardly play about with as you can if it only weighs 10 cwt., and it usually forms the base on which the boat is built, being in fact the first instead of the last operation. Nevertheless, it is suggested that even in a yacht of this class it is well worth while to weigh all parts, at any rate from the designer's point of view, as he can gradually build up a most useful file from which he can estimate the C.G. of any yacht he may be designing with considerable accuracy. I believe it is usual in yachts with only part ballast outside, to assume the C.G. to be amidships, excluding special heavy weights such as mast, engine, etc. Moments are taken about the C.B. for these, and their resultant considered to be the C.G. of the yacht. If this is found to be any appreciable distance from the C.B., the C.G. of the keel is adjusted accordingly, the exact trim being arrived at by placing the inside ballast as necessary. This, of course, has its points, as it is not necessary to be so precise as it is when there is no inside ballast to help to find the correct trim; it is also possible to experiment with different trims, but it emphasizes the point that in some ways far more care is required in designing a small yacht with all outside ballast. This chapter has been written with a view to helping in this admittedly rather difficult matter, by a method that the writer has proved sound, and it is hoped that it will be found of use in attaining the desired end.

CHAPTER XIII

Odds and Ends

THE TABLE OF OFF-SETS

BEFORE a yacht can be built she must be *laid-off*, or as the Americans term it, *lofted*. A full size drawing is made on a lofting floor, which is a flat floor large enough to take the full sized drawing. It has a matt surface and is coloured black. The long construction lines such as the L.W.L. are made by stretching a bit of chalked string and twanging it. Then using an appropriate scale, say 1 in. to a foot, the measurements of the body-plan are transferred to the floor. Fig. 32 is a drawing of a yacht that I designed in 1911. She is 18 ft. on the L.W.L. with a beam of 6 ft. and a draught of 3 ft. 6 in. The drawing was published in the *Yachting Monthly*, Vol. XI, page 54, together with the table of off-sets. Some time after the publication of the lines of *Memory*, as she came to be called, I received a letter from my friend, Mr. H. J. Suffling, inviting me to go to Yarmouth and sail in a yacht that he had had built by Mr. Ernest Woods of Horning. I accepted the invitation and had several enjoyable and instructive cruises in *Memory*. She was a beautifully built craft, and although she obviously had a very poor metacentric analysis, she was perfectly docile, a delightful little ship in every way. Had she been Bermudan rigged she would have probably shown vices, for as I have already said it is a mistake to fit an unbalanced yacht with this rig. Mr. Suffling gave Woods the table of off-sets and the drawing in the magazine, and this was quite enough for the purpose. Three of my designs, one of them *Sinah*, have been built as though the body-plan sections represented the inner aspect of the planking instead of the outer surface. In olden days it was the custom to draw the designs in this way. Another mistake which I am sure, judging by the appearance of the sterns, has been made in some of my transom stern designs has been the failure to allow for the thickness of the planking in cutting the transom, although it has been allowed for in the sections. These yachts appear to have a hollow aft in their top-sides, as indeed they have. To avoid these mistakes it is best to furnish the builder with a table of off-sets, although if he has a suitable scale he can make one himself, or take off the measurements. If a drawing scaled to inches is

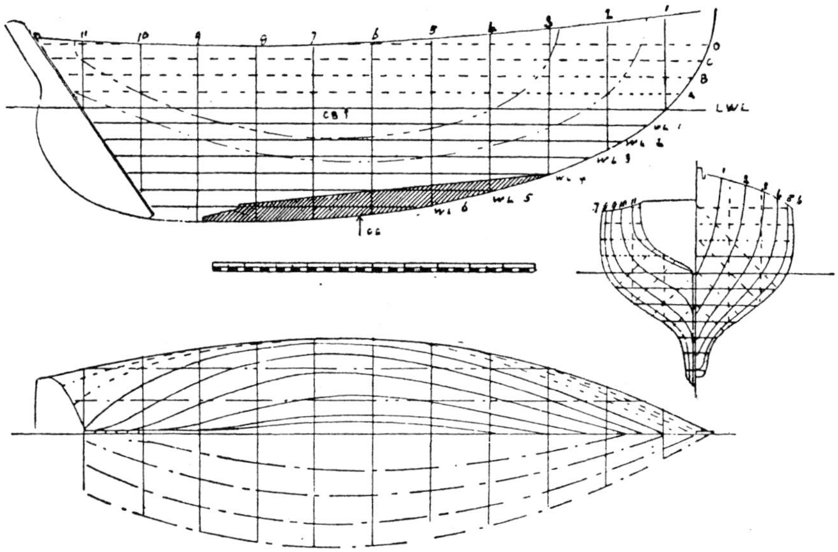

Fig. 32. *MEMORY*, designed and built in 1911, had a very poor metacentric analysis, but was perfectly docile.

being built to by a Continental builder he will take off the measurements from the drawing with a metric rule and multiply each reading by 12 or 16 if the scale is 1 in. or $\frac{3}{4}$ in. to the foot. It is quite simple, but tedious to construct the table. We first measure from the sheer-plan the heights of each section from water-line to rail. Then the depths to lower aspect of the keel below the L.W.L. Then we measure the width of the deck at each section, and finally all the water-line widths from the L.W.L. downwards, and the level line widths from the L.W.L. upwards. Then make a note of the angle of the stern post, the length of the forward and after overhangs and the spacing of the sections and water-lines. If we want to be very accurate we can measure up the diagonals and the bow and buttock lines. Then we can cable figures to Sydney and our friend there can at once start on his yacht.

The Curve of Versed Sines and the Trochoid

In the design of *Sinah* on page 42 the reader will have noted that within the curve of upright areas a circle has been inscribed and two more curves drawn. The forward curve is known to mathematicians as *a Curve of versed sines*. It will be noted that the forward and after part of the curves are identical but reversed. The after curve is a *trochoid*, the curve that would be traced by a pencil attached to the circumference of the circle

PLATE 9

Merrythought (19′ LWL), Zyklon design. Built 1938 in Plymouth for C.F.D. Roberts. She is now in Scotland. Alternative sail plan has no bowsprit. Not a Z.4 tonner.

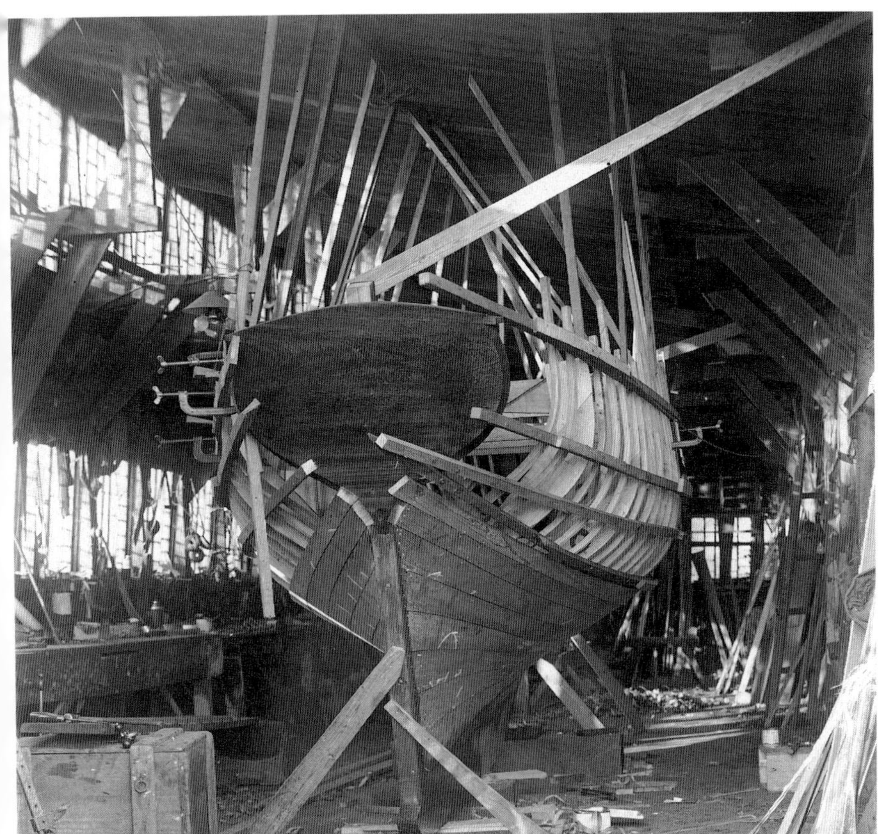

Grey owl under construction.

PLATE 10 *Vindilis* (22′ 6″ LWL). Davinka/Vindilis design. Built 1935 by A.H. Moody & Son for THB and launched by EB. She cruised from the Solent to the West Country, Channel Isles and Rouen; she is now in Scotland. (above left) *Vindilis* under power, OJB, RDWB, THB. (above right) The table in use, l to r ENB, THB, EB. (below) Interior, looking aft. Note the offset companion; the table is under the carpet. The photograph shows some of her 31 lockers (and their contents). Note the sense of space.

when it is rolled from its central position to the after end of the base-line. The construction of two curves is obvious from an inspection of the drawing. The generating circle is placed at a spot 11/20ths of the L.W.L. from the fore end. We divide its circumference by ruling a line parallel to the base-line through the circle; another vertically through the centre, and two other lines at an angle of 45 degrees to these lines. Divide each part of the base-line, that before and that behind the circle, into four parts. Rule three lines parallel to the base-line through, first the centre of the circle, and then through the points where the diagonal lines cut the circumference. In the fore half erect three perpendiculars through the dividing points of the four sections. These lines will cut the horizontal lines at three points. Join the top of the diameter with the three points and with the fore end of the L.W.L. The resulting curve is the curve of versed sines. Now turn to the after end; join the bottom of the semi-diameter with the spot where the middle horizontal line cuts the circumference and also with the spots where the upper and lower horizontal line cuts the circumference. We have three lines radiating fanwise from the bottom of the semi-diameter. Draw a line parallel to these three lines from the after dividing point of the four sections of the base-line, from the middle point, and from the upper point, the 45-degree point. Check these lines by measuring along the base-line and from the upper ends of the fan-radii. Join up the points of intersection with a curve; this is the required trochoid. Theoretically these curves should coincide with the curve of areas. It is usual to prolong the ends of the curves beyond the ends of the base-line so as to eliminate the sharp end; the first bit of faking. The position of the generating circle is empirical, and worse than all the trochoid depends entirely upon the relationship of the diameter of the generating circle with the base-line. By altering the scale of the ordinates one can get a close approximation. Dixon Kemp gives a formula for discovering the proper relationship and Skene has some factors. But very fortunately I have just discovered that Dixon Kemp says that for all practical purposes a curve of versed sines is as good as a trochoid for the after-body. Now what does this all mean? Probably little or nothing! In 1877 Colin Archer propounded his wave-form theory which was a modification of the Scott Russell theory. We need not go into this theory; suffice it to say that a wave has a cycloidal form, and the disposition of the displacement as shown by the curves just described will harmonize with it. As far as I can judge from the curves that I have made, it would appear that the fore and after body should both be a little fuller than the curve of versed sines. The correspondence probably has no connection with wave forms, but it is a good guide. If the displacement curve were definitely fuller than the curve of versed sines, I should think that it was too great and *vice versa*.

Improving a Design

There are very few designs that are not capable of improvement. When I have finished one I am generally rather pleased with it. In a few months I see faults, and soon I feel that the whole thing must be redrawn. The metacentric analysis has made yacht designing more complicated; it has shown that the fore-body must harmonize with the after-body and that it is impossible to dissociate the hull proper from the keel appendage. A hull may be balanced in two ways: intrinsically and extrinsically. If the hull proper is balanced metacentrically and by coalescence of H.B. and U.B., then when we add a fin, or a keel appendage, this must be centrally placed. I call this intrinsic balance and it is rarely attained. If on the other hand the hull is only partially balanced and the final balance is attained only by the fin being placed before or behind the central spot, actually in almost every case placed well aft, then the hull is extrinsically balanced. One would imagine that intrinsic balance should be aimed at, but this must be a matter for research, and in any case this balance is almost impossible to attain in a small cabin cruiser, but can be managed in a day boat or a larger cruiser. In discussing *Sinah* earlier in the book I have suggested that she would be better were the deadwood aft cut down by bringing the rudder forward and increasing the rake of the stern post. One might be apt to think that it was an easy matter; just draw the keel as you want it! But the keel alteration is interlocked with the metacentric shelf, and it will be found that some modification will be necessary in the bow and stern sections. If we cut down the after deadwood then the weather helm ought to be increased; I do not say that it will be, because ships are funny things. This demands that the bow sections shall be given more swell in proportion to the stern sections, or that the reverse process be adopted, or a combination of both. A few trials, and we shall discover what alterations are necessary to balance the modification of the underwater profile.

In Chapter XI I pointed out that when a metacentric analysis had been completed and was obviously unsatisfactory, alterations in the hull form would be necessary. How shall we set about it? Let us take a concrete example. In 1935, just before I had grasped the metacentric theory, I designed *Aristene* and she was built in Australia at Adelaide. I have never been able to find out anything about her actual sailing balance. I had for

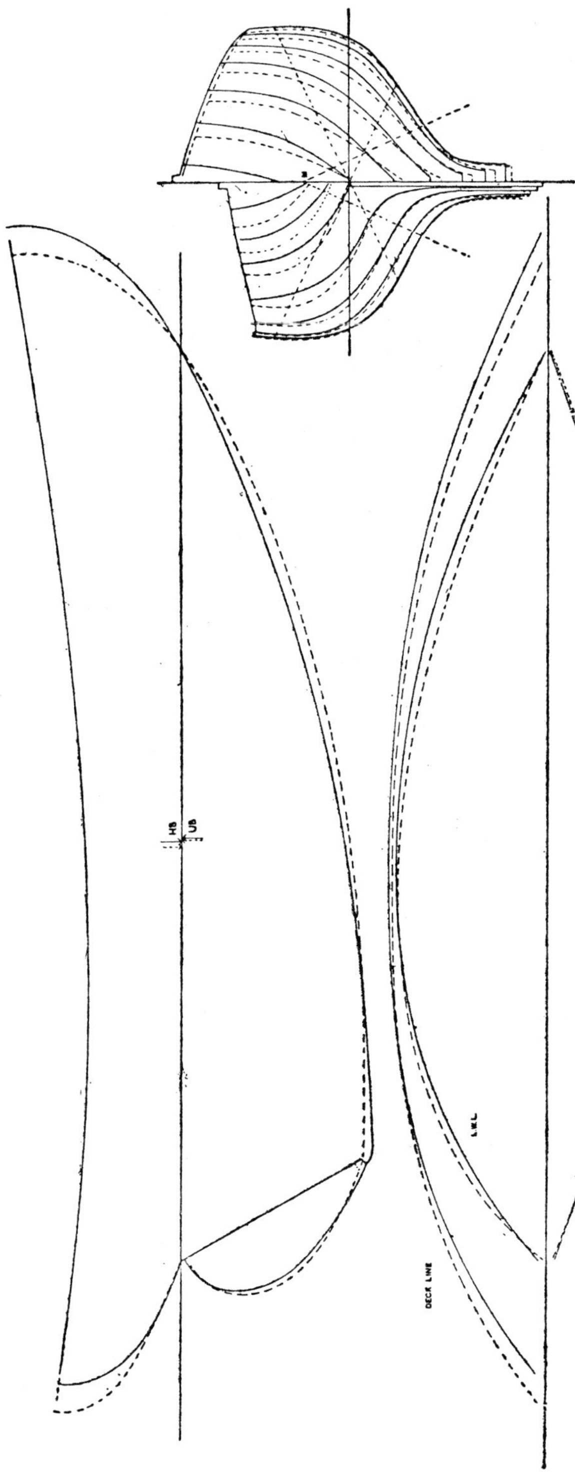

Fig. 33. Improving a design. The original *ARISTENE* is shown by pecked lines, while the improved design, *QUEEN-OF-ARDEN*, is indicated by continuous lines.

years been trying to achieve balance, and in my later designs I got very near it, so *Aristene* ought to have been pretty good, especially as she had a canoe stern that makes balance easy to attain, but when later on I came to analyse her I discovered to my horror that she had a crossed shelf. This fact shows how necessary it is to have some theory to help the eye. The alterations I made to correct the defects are shown in Fig. 33; the changes in the sheer-plan are slight. The canoe stern has been made shorter to bring in the deck line aft. The fore-foot has been somewhat cut away, and the forward overhang increased. The body-plan shows that the bow sections have been swelled out and the quarters well fined up. The alteration in the shelf is noteworthy. Originally crossed, it is now symmetrically disposed fore and aft. The curves of analysis are good, $a=c$. But I am not yet satisfied; I now want to cut away the after dead-wood; this will mean still more fullness in the bow, and she will be better for it. The design of *Queen-of-Arden*, the improved *Aristene*, was published in *The Yachting World*, Vol. X.

I wish to emphasize the fact that the metacentric theory has no necessary connection with displacement, with speed, or even with sea-kindliness. In a recent number of a yachting journal I read in the description of an exceedingly well thought out design the following statement: "It would be easy enough to make the design conform to the metacentric theory, but I prefer the balance just like it is for my own personal viewpoint. I think to conform to the theory would mean alterations in the lines which to my mind would make her possibly *slower to windward*, and make her *too lively*, and so *less comfortable*." The theory has nothing to do with these matters. I could have taken *Aristene* and altered her bow only. This would have increased her displacement and made her less lively. I could have left her bow alone and fined in her stern still more. This would have given her less displacement, and made her perhaps a diver. I cannot see how adherence to the theory could make a yacht slower to windward. We know only a little about hull balance, but there is no doubt that many yachts with a poor analysis are good performers, and quite docile. On the other hand I have only heard of one yacht said to be balanced that had unpleasant habits. If you balance you play for safety, and as far as I can see, take no kind of risk. All the designs in this book are balanced and I doubt if the ordinary yachtsman would grasp the fact that they were otherwise than quite ordinary. In fact in many cases it is impossible to know that the yacht is balanced from a study of the lines without an analysis. Of course an expert can make a very shrewd guess, but he may be wrong.

How to Draw a Chain

Often in drawing a cabin-plan we wish to indicate the position of the anchor chain. This may at first sight appear to be a difficult bit of work; in actual fact it is quite easy.

Begin by drawing four parallel equidistant lines. They are separated by the thickness of a link. The secret is this: *ink-in the two middle lines at once.* Now draw one link of the length desired. This can be obtained by direct measurement from an actual chain, but in our work such accuracy is unnecessary, in fact not desirable, for the chain must be drawn abnormally

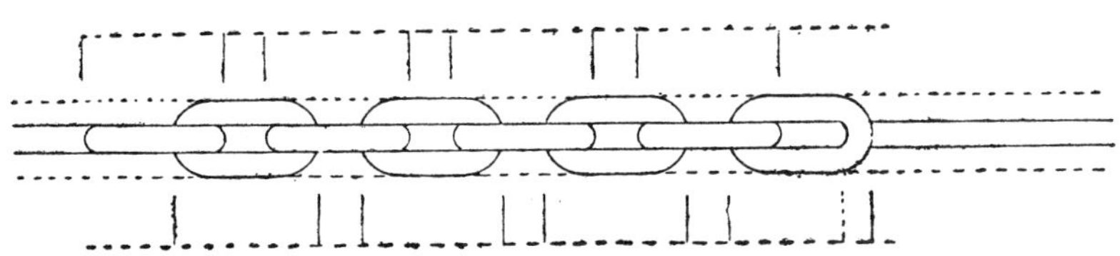

Fig. 34. How to draw a chain.

large if the design is to be reproduced. If it is too small the links will run together in the printing and the effect be spoiled. Having drawn one link, then cross it with another. Remember that the place of crossing will be twice the thickness of the individual link. Measure the second link to see that the two are of equal length. Having got two you have got your measurements and can carry on without difficulty. In the ordinary cabin-plan most of the work is done freehand with a mapping pen. The method is made quite plain in Fig. 34.

Lifting Cabin-tops

The sail-plan of *Paida* on page 69 shows a lifting cabin-top instead of the built-up top-sides seen in the line drawing on page 50. Sleeping accommodation can be obtained in a small yacht or boat either by a tent, by a lifting top, or by a fixed deck. Lifting cabin-tops are very common on the Broads' yachts, but they in general lift in one sense only: the front is hinged and the back rises. This of course gives added headroom at the after end of the cabin only. At the end of the year 1900 *The Yachtsman* staged a competition for a lifting cabin-top. None of the solutions sent in were satisfactory, especially for a yacht of the canoe-yawl type, but on January 31st, 1901, a thoroughly workable plan devised by the late Mr. Umfreville Lawes under the pseudonym of A.D. 1900 was published in *The Yachtsman*.

The method is shown in the accompanying drawings and is described by its inventor as follows: "It will be seen that the cabin-top lifts in two sections both forward and aft. The coamings are in two parts, the upper part being strongly hinged at the fore end of the cabin-top deck and at the after end of the lower part. Figs. A, B and C show the hinges. Fig. B shows the cabin-top lowered; Fig. A shows it raised. The lower coamings are bound at the forward end with brass angles, which project about two inches above to receive the upper coamings when lowered, and so prevent any lateral play. The cabin sides consist of light duck painted with Berthon flexible paint, and in them may be inserted if desired windows

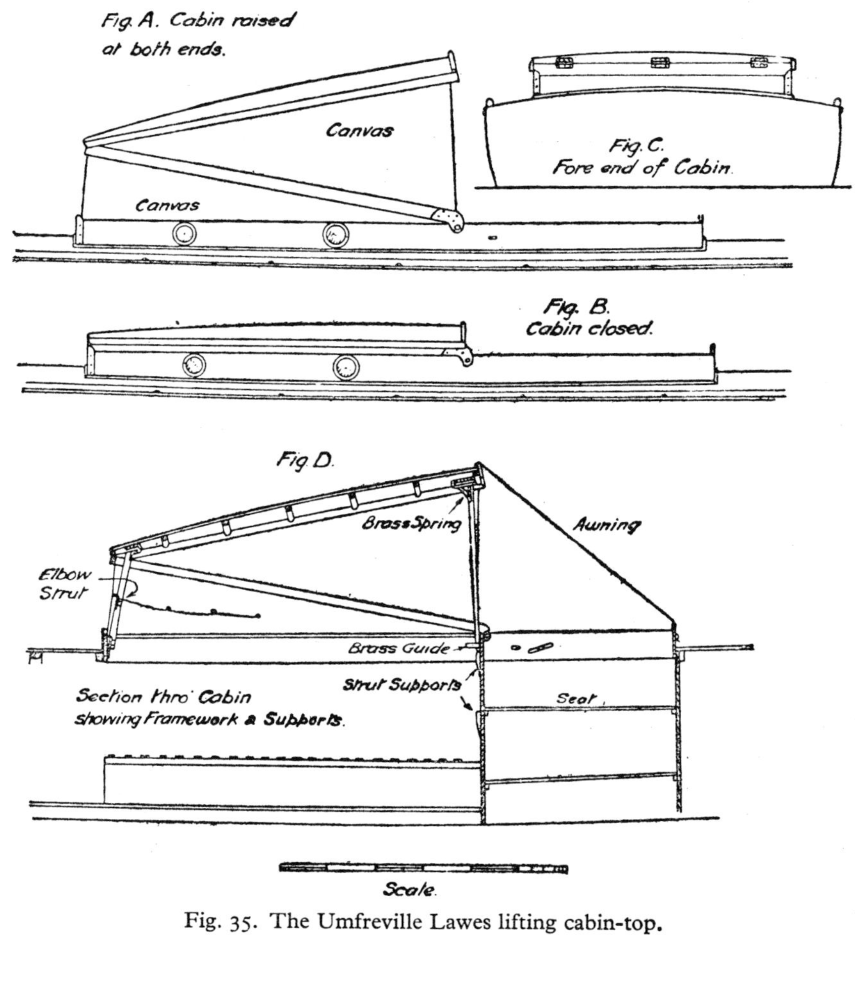

Fig. 35. The Umfreville Lawes lifting cabin-top.

made of celluloid. Although the arrangement looks flimsy and fragile when raised, it is in reality very firm and will easily bear the weight of a man on the top when raised or lowered. The supports adjust themselves automatically, the forward ones being on the elbow joint principle, and the after ones having fixed behind them near the hinges a strip of hard spring brass, which forces them against the bulkhead, so that they cannot fail to engage on the rests fixed thereon. The hinges must be very strong. Fig. D shows the struts and rests. In order to prevent the canvas at the fore end bulging out when in the act of lowering, several small brass rings are sewn on to the sides and fore end (see Fig. D) and through these a small cord is rove and attached to the elbow struts, which as they close tauten the cords, and thereby pull the slack of the canvas inwards clear of the coamings. The after end of the raised cabin may be closed by a light awning stretched from the after end of the coaming (Fig. D) and buttoned or hooked round the sides. Should the wind be very heavy, half only of the cabin may be raised, and if the yacht is lying to the tide with the wind aft, causing her to sheer about, the after end may be lowered and the forward end only kept up, and vice versa."

My late friend, L. Boughton Chatwin, bought an Iona fishing boat, 18 ft. long, and fitted one of these lifting tops. In her he and his wife did some very extensive cruising, including a somewhat hazardous crossing of Poole Bay in an easterly breeze. She was at one time a familiar sight on the Hamble, and then she migrated to the Midland Sailing Club's Lake, near Birmingham. This is the best form of raising top that I have seen, and is far superior to a tent.

Scantlings and Specifications

I am constantly being asked for the specifications of yachts that I have designed, so I am including those of a 4-tonner and an 8-tonner. The scantlings of *Thuella* and *Aella* were used for one of the *Zyklons* (the Z 4-tonners) that was built at Cremyl. I saw the yacht completed and she seemed a most satisfactory job.

In this particular yacht the stern knee and stem were bolted to the keel with galvanized steel bolts, which seems to me to be quite unobjectionable and makes for strength and economy. In no case must naval brass be allowed to come into contact with steel floors. Some years ago I had a letter from Olin Stephens asking me how our builders managed to avoid galvanic action when they bolted steel floors with brass bolts. My answer was simple: "They cannot avoid it." I know of one yacht today whose floors have corroded away for this reason. One cannot defy natural laws, and if two metals of different electric potential are placed in contact in an electrolyte (in our case sea water) an electric current will start, and the more electro-positive metal will be dissolved. The usual metals used in yachts are zinc, iron, lead, tin and copper. These are arranged in the order of their electric potentials, zinc being the most positive, and copper the most negative. Thus a copper-containing propeller will act upon a ship's plate containing iron or upon a steel A-bracket of the propeller. To avoid this action it is usual to place zinc plates near the brass structures so that the zinc may waste away and the iron be spared. It is obvious that if we place a zinc galvanized surface in electric contact with a copper-containing bolt, the galvanizing will rapidly waste away, and then the iron will follow. Yet when *Vindilis* was built I had the very greatest difficulty in persuading her builders to use steel bolts for her floors. It is the usual custom in this country to rig up a battery in the bilges that will inevitably destroy zinc and iron. A lead keel may be bolted with ungalvanized iron bolts because in the first place the electric potential between the two metals is very similar, and in the second place the lead keel will be eaten away and the iron bolt spared, and as there is plenty of lead to be dissolved it does not matter. If you must bolt the lead keel to an iron floor use iron bolts. But why not spend a few more pounds and have Dixtrudo forging-bronze floors, and use naval brass bolts and so avoid galvanic actions? In the case of *Vindilis* the increase in cost would have been £8 10s., but I did not know this at the time, so *Vindilis* has steel floors with steel bolts, whereas *Lindy II*, which was built on the same moulds, had Dixtrudo.

The specification given for *Dream-of-Arden* is that to which *Vindilis* was built, though I have made a few alterations to bring the weight down without any real loss of strength. These specifications are far heavier than many designers use. In *Yachting World* for April, 1944, and in the subsequent numbers, there will be found the scantlings of designs submitted for a designing competition for a fast cruiser of 25 ft. on the L.W.L. These are all far lighter than those that I have adopted, so it is possible that we might split the difference. In any case these scantlings must be regarded as the absolute maximum for an 8-tonner.

In both cases the yachts are built entirely on bent timbers. This has been the practice in all yachts up to about 8 tons that have been built to my design, and as far as I know this method has proved perfectly satisfactory.

The late Claud Worth used to say that yachts ought to be built "all-grown" or "all-steamed," and that the mixture was bad. His view was that if the hull got a squeeze the all-steamed type would give and no harm would be done, and that the all-grown timbered yacht would be strong enough to withstand the strain. The mixed hull would "fall between two stools," to mix our metaphors, and some of the grown timbers might fracture. Be this as it may, the common practice in larger craft is to mix the two methods. In *Khamseen* I have introduced one bent frame between each grown frame, and the frames amidships are doubled. This construction was taken from a design by the late Albert Strange. If the frames are all grown the fastenings may be iron dumps instead of copper, which is of course cheaper and stronger, and the iron lasts as long as the timbers. Shortly before the war I saw some dumps that had been in a hull for forty years, and they were quite bright. There used to be a most beautifully built yawl of about 40 tons, called *Armorel II*. Her owner was a barge owner and she was entirely iron fastened. Mr. Peters told me that he had had this done because of the increased strength, and he said that the iron fastenings of a barge lasted as long as the hull.

SCANTLINGS FOR *THUELLA* AND *AELLA*

Stem, stern post and stern knee: Oak or elm, sided 3½ in.

Transom of "Aella": Teak, 1½ in.

Planking: Pitch pine, finished, $\frac{13}{16}$ in.

Garboards: Elm, ⅞ in. fastened with 11 gauge copper nails, clenched on roves.

Timbers: Steamed American elm, 1⅜ in. by 1 in. spaced 6 in. centre to centre.

Stringers: 2½ in. by 1½ in., Oregon or pitch pine.

Shelf: 3½ in. by 2 in.

Deck: ⅞ in. pine, tongued and grooved and canvas covered.

Cabin deck: ¾ in. pine, tongued and grooved and canvas covered.

Coamings and sheer-strake: ¾ in. teak.

Beams: Oak or ash or wych elm, 1½ in. by 2½ in.

Main beams: 2¼ in. by 3 in.

Floors: Galvanized iron, 1¼ in. by ⅜ in., iron bolts ¾ in. No brass bolts through galvanized iron floors.

Carlines: 2½ in. by 2 in.

Mast: If solid, heel 3¾ in., deck 4¾ in., very gradual taper to 3 in. at sheave.

Anchor: Fisherman, 30 lb.

Chain: $\frac{5}{16}$ in.

Warps: Two 15 fathoms, 2½ in.

Lead keel: ⅞ in. brass bolts not through floors.

Iron keel: ⅞ in. iron bolts quenched in oil.

SPECIFICATION OF *DREAM-OF-ARDEN*

Fastenings: To be of copper or brass riveted over roves with the exception of centre line, where the fastenings are to be of galvanized steel nuts and bolts. The lead keel is to be bolted on with inch bolts of Naval brass. These bolts are not to come in contact with the iron floors. Stopwaters to be fitted where necessary. Keel bolts to be staggered alternately port and starboard.

Planking: Pitch pine, 1 in. All planks in one length $\frac{7}{8}$ in. finished and no stealers. Garboard 1 in. finished.

Keel: English oak, sided $4\frac{1}{2}$ in. moulded as in drawing.

Stem: Natural grown English oak, $4\frac{1}{2}$ in. in one piece if possible.

Stern post: English oak, $4\frac{1}{2}$ in. tapered aft to match fore side of rudder. Aft end of stern post hollowed for rudder.

Frames: American elm, oak or ash, $1\frac{1}{2}$ in. by $1\frac{1}{4}$ in. bent. Spaced 6 in. centres, an extra frame to be worked in the way of the mainmast.

Floors: Galvanized wrought iron.

Shelf: Pitch pine in one length, $1\frac{1}{2}$ in. by 5 in. Top of shelf to come right up to deck.

Bilge and side stringers: Pitch pine in one length, $1\frac{1}{2}$ in. by 2 in. Shelf and stringers to be fastened to alternate timbers.

Deck: Pine, tongued and grooved, covered with canvas, laid 1 in. narrow planks.

Cabin-top deck: Pine, $\frac{3}{4}$ in., tongued and grooved, covered with canvas, with 2 in. teak margins. Teak handrails outside and inside cabin.

Coamings of cabin-top and cockpit: Teak, cabin-top 1 in., cockpit $\frac{3}{4}$ in. American elm capping round cockpit.

Deck beams: English oak, 2 in. by $2\frac{1}{8}$ in., spaced as in drawing about 14 in. to 15 in., to be dovetailed into shelf and fastened with stout brass screws.

Main beams: English oak, $2\frac{3}{8}$ in. by $2\frac{7}{8}$ in.

Hanging knees: Galvanized iron straps or hanging knees, $1\frac{1}{2}$ in. by $\frac{3}{8}$ in., to be fitted to six beams each side.

Cabin-top beams: English oak, $1\frac{1}{2}$ in. by $1\frac{1}{2}$ in.

Breasthook: English oak grown, or steel.

Deadwood and after knee: English oak, 4 in.

Lodging knees: Grown English oak to be fitted as shown in drawing, or steel.

Rudder: English oak, 3 in. forward tapering to $1\frac{1}{2}$ in. aft, to be attached to stern post by brass strap and to keel aft by iron shoe. The shoe to be provided with a pin to fit into rudder.

Rudder stock: Galvanized steel with forks as shown in drawing.

Tiller: Galvanized iron or English oak.

Bulwarks: Teak, 1 in.; 4 in. high forward tapering to $2\frac{1}{2}$ in. aft. To be fitted with American elm capping.

Cabin bulkheads: Plywood.

Deck fittings: All deck fittings to be of best galvanized iron. Fairleads of *lignum vitae* with straps through beams.

Navel pipe: Galvanized iron and all other galvanized iron fittings for deck, mast, etc., to be supplied and fitted.

Stout galvanized iron horse for mainsheet, and *horse 2 ft. long* to be fitted through deck beams for foresail sheet.

Galvanized iron pinrail round mast. The forward bolt to be connected to a through bolt in the keel by a stirrup chain and rigging screw.

Samson post of English oak to be fitted forward and two *oak bollards* aft. Bollards to have brass pins.

Winch: Simpson Lawrence Junior oil-bath.

Lightboards to be made and fitted to rigging.

Chainplates to be made of galvanized wrought steel fitted outside and through fastened to frames.

40 *lb. fisherman type anchor* with long shank or *C.Q.R.* and 28 *lb. kedge.*

30 *fathoms* $\frac{3}{4}$ *in. short linked galvanized chain,* correctly pitched to fit gypsy on winch (Connop Bros.), tested.

Two warps, each 15 fathoms of 2 in. coir rope.

Three portlights a side and one forward of cabin-top, all to open. Each portlight to have outside rim covered by turned teak circle.

Gammon iron to be fitted with a roller on each side and a sheave on the top so that fore stay can be set up to Samson post.

Boathook.

Ballast: Lead keel about 2 tons as in design, to be bolted as previously described. Inside ballast to be of lead in small pigs sufficient to bring yacht to designed water-line.

Interior: The interior is to be fitted out as shown in cabin-plan, the galley being aft on starboard side. Bunk seats to be arranged on port and starboard sides with padded backs in two parts hinged to come forward to expose bedding lockers.

Mattresses and cushions to be supplied by owner.

Cabin table to fold into floor as shown in drawing.

Teak companion with rubber treads from cabin to cockpit.

Floor boards of oiled teak, $\frac{3}{4}$ in.

Water tanks to be fitted under seats and pump to be supplied near galley.

Galley to be lined with approved material (Monel metal, Formica).

Forecastle: forepeak to be arranged as cupboard with shelves and two doors.

Forecastle to be fitted with Baby Blake W.C. and wash basin.

Two canvas cots with galvanized iron frames to be fitted as shown and to be laced as described.

Chain locker to be arranged as in drawing.

Cockpit: To be arranged as shown with sail locker aft, larder and lockers. Ventilating holes in larder to be protected by perforated zinc.

Cockpit benches to be of $\frac{1}{4}$ in. teak.

Bilge pump: force pump type discharging through top-sides, to be fitted with strum box which can be easily disconnected.

American elm grating to be let into centre of cockpit floor.

Running and standing rigging: To be as in plan.

Auxiliary engine: Stuart Turner 8 h.p. with reverse but no reducing gear.

Sails: To be provided by builder. To be bent and fitted by builder. Builder to supply and fit all spars.

Spars (as in drawing): Main to be fitted with Laurent Giles Birmabright track and slides.

Canvas covers: Builder to supply canvas awnings for fore hatch and cockpit and canvas cover for winch.

Forehatch: To be fitted with deadlight and side triangular flaps, to be fitted with long hinges.

Decklights: Two prismatic decklights to be fitted in deck in forecastle and one over sail locker aft.

Painting: Top-sides to be painted with best black enamel with gold line, sunk in a cove.

Bottom and boot top Kobe green.

Interior, best white enamel on under side of deck between beams; forecastle varnished.

Insurance: The yacht is to be fully insured during building and on trials.

Registry: The Board of Trade markings to be carved on beams as required.

CHAPTER XIV

A Selection of Designs

AELLA

L.O.A.	23 ft. 6 in.
L.W.L.	20 ft.
Beam	7 ft. 6 in.
Draught	2 ft. 8 in.
Draught with centre-board	4 ft. 6 in.
Displacement	3.1 tons

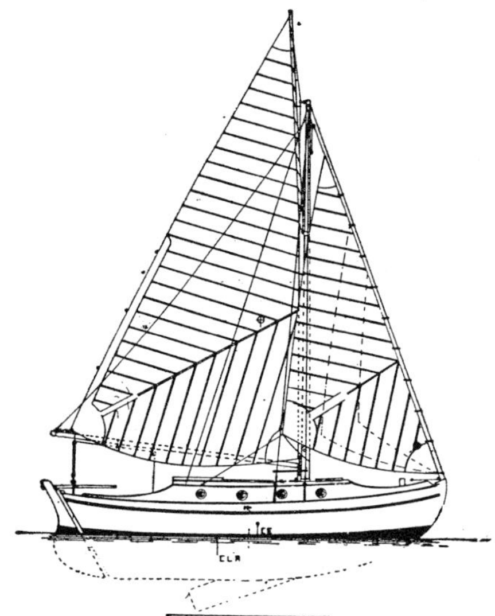

I WAS asked by a Midland yachtsman to design him a yacht suitable for cruising on the Trent. The draught had to be kept low, and after some discussion, he wanting less and I more draught, we decided upon 2 ft. 8 in. The yacht was built at Woodbridge and called *Rami II*. Unfortunately her owner decided to dispense with her centre-board, so I have no knowledge of how she would have behaved with a board. In any case he was well satisfied with her, and said that she went to windward quite well. I thought that she was a very nice boat in those distant days, but now I know that she has a crossed shelf and is capable of improvement.

When the editor of *The Yachtsman* discussed with me a suitable subject for his next designing competition, I suggested that *Rami II* might be a good type, and she was adopted. When I was asked to be one of the judges I thought that I would bring *Rami II* up to date, and *Aella* is the result. *Aella* is the poetic form of *thuella*, the Greek for a storm, especially of the whirlwind type.

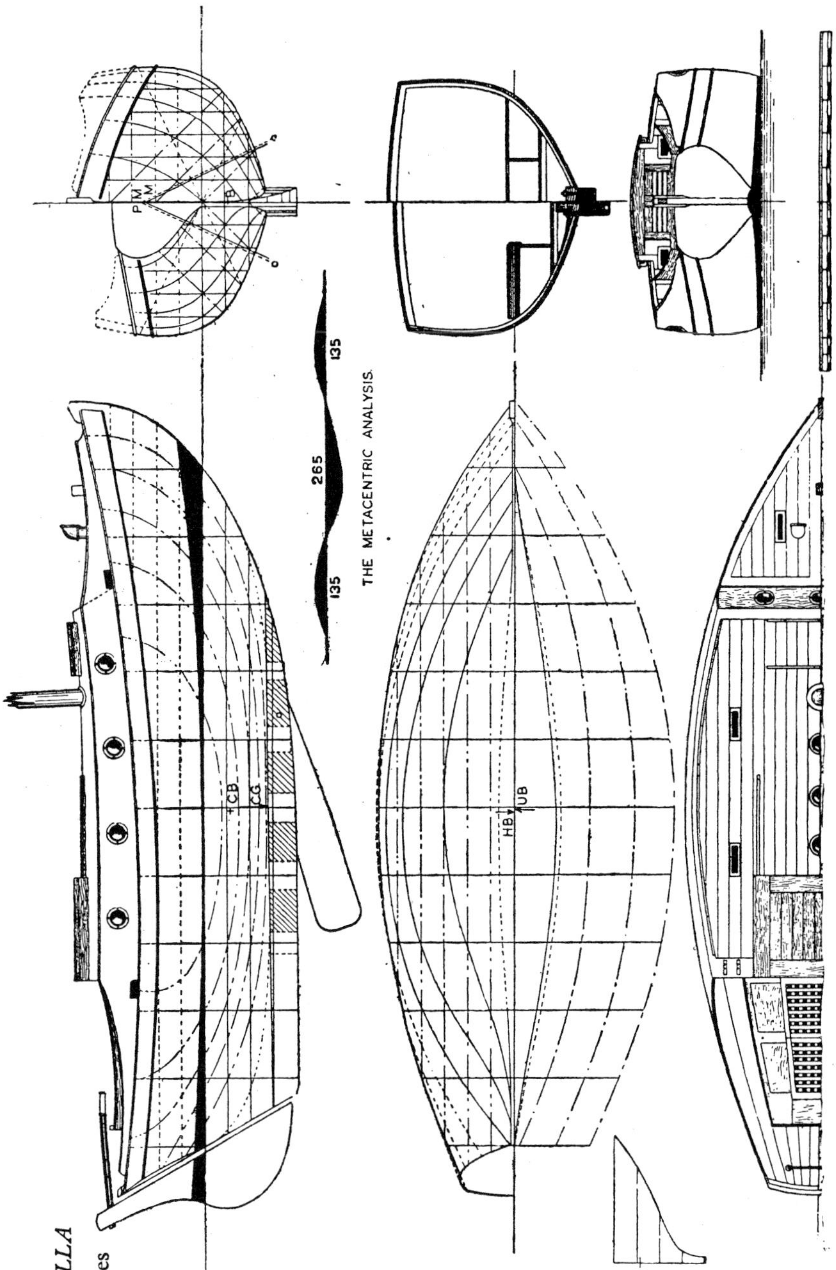

AELLA
Lines

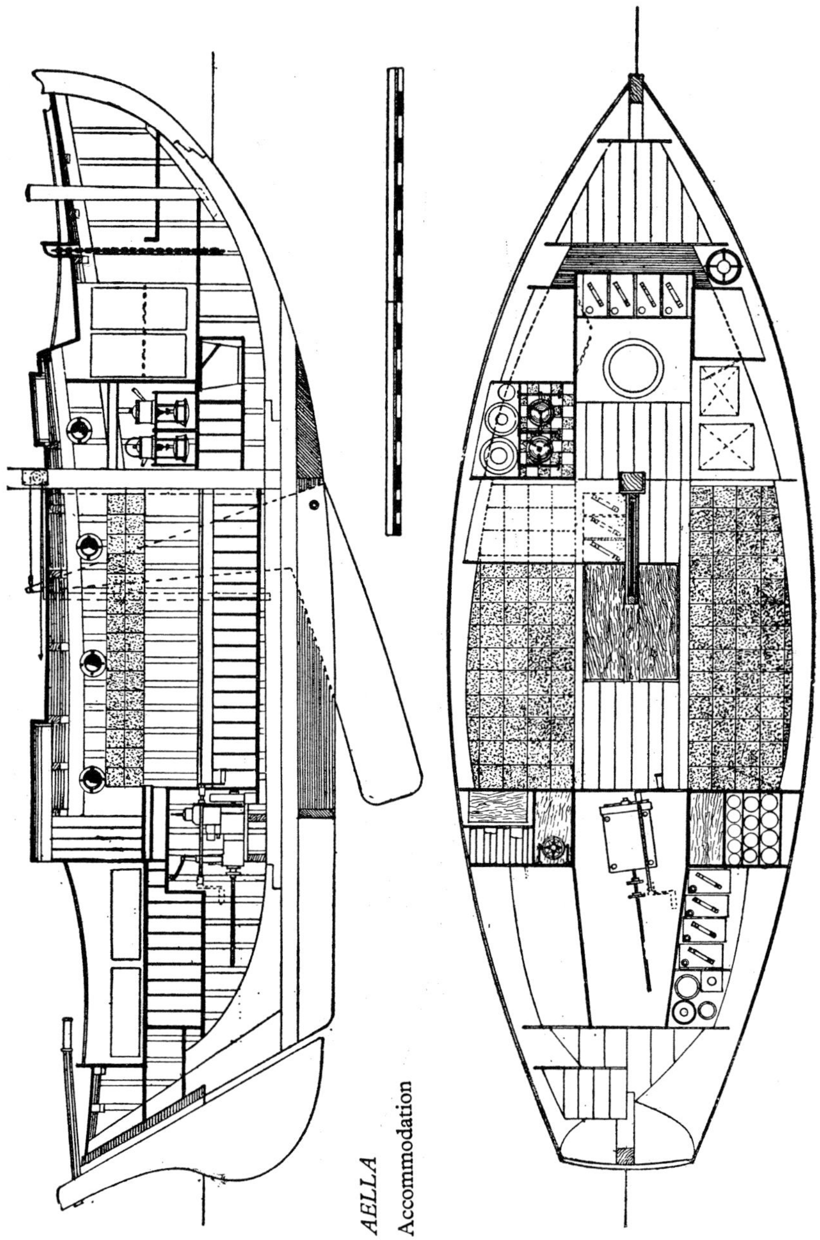

AELLA
Accommodation

Aella is designed for service in shallow waters such as the East Coast rivers, the Thames Estuary and similar waters. She would be suitable for amateur construction; she has an ample range of stability and it would be very difficult to capsize her. The lines give a perfect metacentric balance, and to achieve this with a transom sterned boat we have to adopt a V-shaped section giving a transom of the Thames barge type; if the upper part is well rolled in the effect is quite good especially if the transom is curved. As I have said before, a curved transom is difficult to construct and adds considerably to the cost of the yacht. I think that the amateur builder had better adopt the flat transom or better use only a slight curvature by planing down a thicker plank than would be used for a flat transom. A curved transom can be made in three ways: It can be carved out of a thick timber; it can be bent; or it can be built. The first method is suitable for a moderate curve only; the second is difficult, for the transom has to be bent over a strong template. This method was used for the transom of *Askadil*, and the builders told me that the curved transom increased the yacht's cost by twenty pounds. Finally the curve can be made by building up the transom with thin planks and doubling them, or by vertical planks built upon a frame; the wide curved transoms of motor yachts are built in this manner. The best plan is to adopt a moderate curve that can be shaped out of timber say twice the thickness of the finished transom.

The lay-out gives comfortable sleeping accommodation for two. The raised top-sides give ample sitting room both in cabin and forecastle. There is a table right across the fore end of the cabin from the centre-board case to the ship's side; in front of this is the two-burner stove in the forecastle. This athwartship table acts as a cooking table, being within easy reach of the galley, and also as a cabin sideboard. There is room under it for the feet when sleeping, or it can be folded back against the ship's side. The main table is attached to the centre-board case. In front of the galley is a seat on the starboard side which is conveniently arranged for the cook; around him are lockers for pots and pans and other galley gear. Amidships is a toilet bucket arranged in a lead-lined locker; in front of it are the water-cans and the chain, while the riding light has its own shelf. In the eyes of the ship is a locker for sails, bags and clothes. Three extra water-cans find a home on the port side of the centre-board case. The centre-board is L-shaped of the Albert Strange-George Holmes type. The lifting arm may be arranged in two ways; it may come out on deck as shown in the cabin plans, or it may be shorter and wholly in the cabin; then the top of the case must be curved and a sliding cover must move with the arm to keep water from spurting out.

A shallow draught yacht loses half its virtues if it is not fitted with a lowering mast. This almost excludes the Bermudan mast, so I have adopted the next best, the gunter-lug. This might well be of the sliding gunter type so ably described by Conor O'Brien.

THUELLA

Thuella is of the Scandinavian type. This form seems to me to have every advantage except one; it is easy to balance, in fact the tendency is to over-balance; the pointed stern is the most seaworthy of all forms; it is easy to build and therefore cheap; the only disadvantage is that the deck space aft is contracted. This, however, is easily overcome; we have saved at least a foot in overall length, so we add it on to the L.W.L. *Thuella* has the same accommodation as the Z 4-tonners that are 19 ft. on the L.W.L. and have transom sterns, and she was designed with a view to easy building and is suitable for amateur construction. The sheer is not excessive and the bilges somewhat slack, which features make for ease in planking. The frame is constructed of members that are almost straight, and the log-keel calls for no adze-work. The bolts are all short and there are no long bores, so difficult for the amateur to perform. The easy bilges may tend to make *Thuella* heel easily at first, but her stability factor shows that she has an ample reserve stability. She ought

THUELLA

L.O.A.	..	..	24 ft.
L.W.L.	..	..	20 ft.
Beam	..	..	7 ft.
Draught	..	..	4 ft.

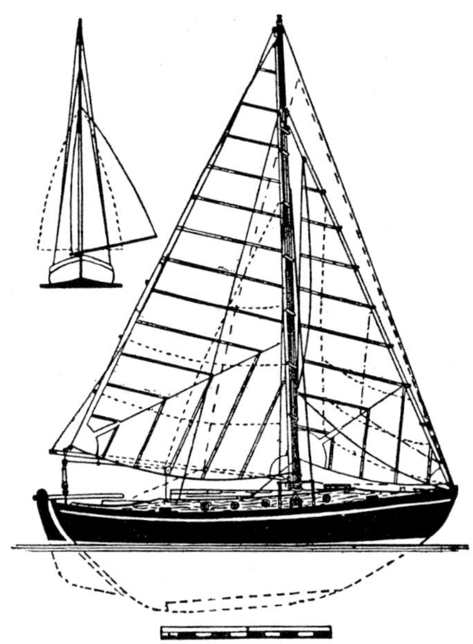

PLATE 11 *Vindilis* (above left) Under sail in the Beaulieu River; (above right) OJB and THB in the English Channel. (below) Interior, looking forward. There is another pipe cot to port, over 'Baby Blake' and locker with wash-basin.

PLATE 12

Vindilis: (above) EB and THB, during their last cruise, in 1939. The 'picture hat' for which EB was teased, was an anti-glare device not merely eccentricity. (left) At anchor in the Medina River, Isle of Wight, as it used to be.

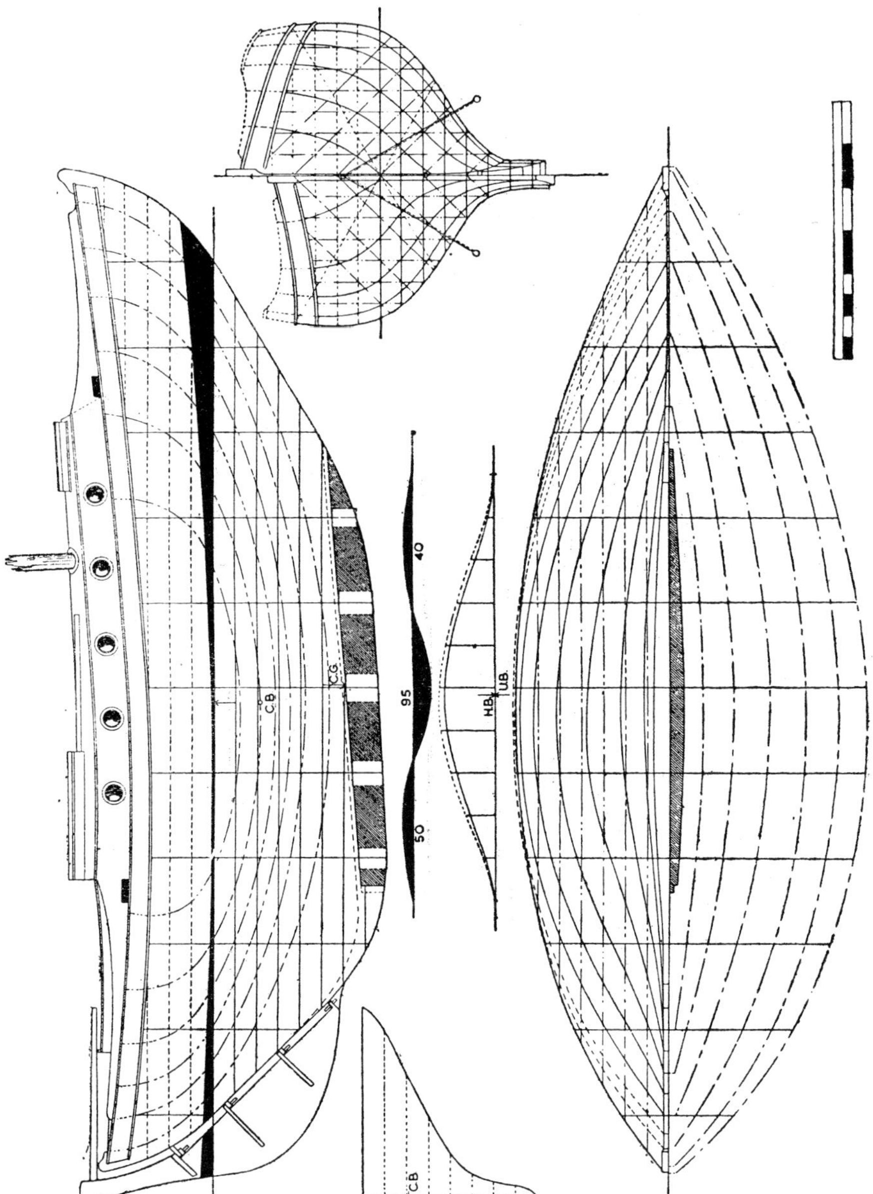

THUELLA
Lines

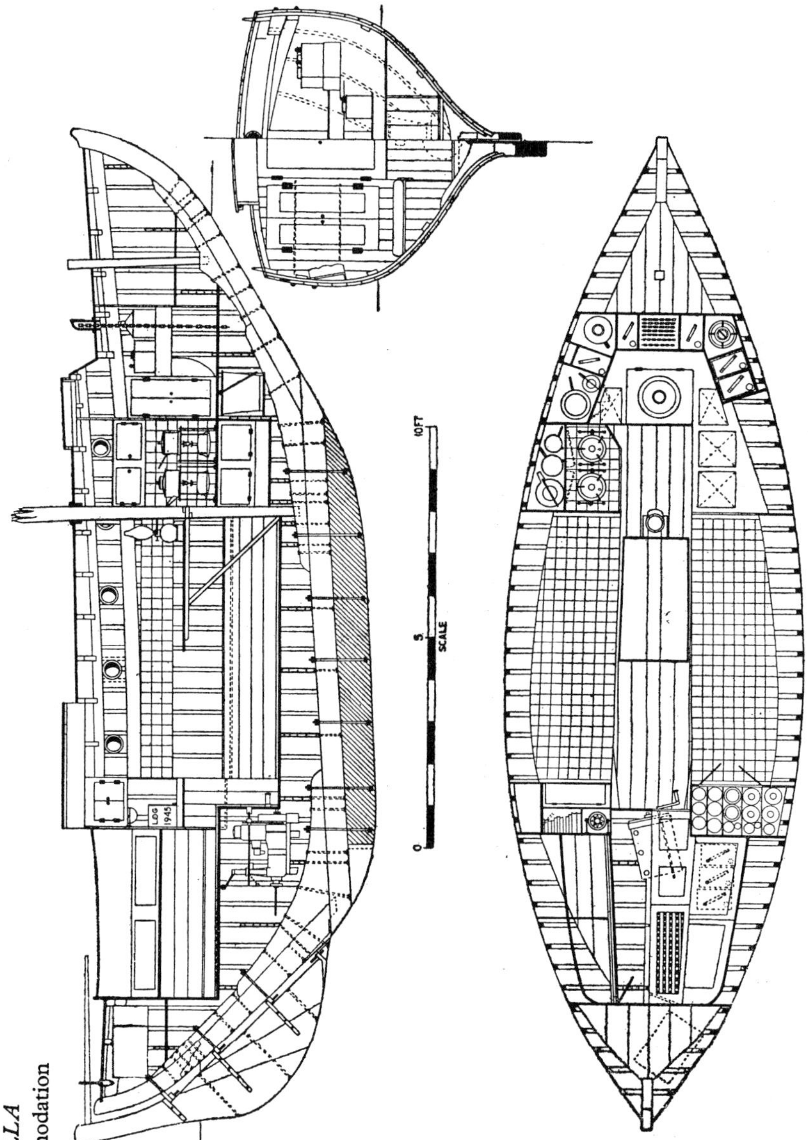

THUELLA
Accommodation

to be very handy, seaworthy and comfortable. The rudder is small and well away from the heel of the keel, but whereas the yacht is perfectly balanced and has a short lateral plane, it ought to be sufficient; in any case the area can easily be increased.

The lay-out is practically the same as that of *Aella* and the drawing is self-explanatory. Two can sleep on the sofas and one on the cabin sole. The raised top-sides amidships, which I call my "centre turret," give ample sitting head-room both in the cabin and forecastle. The break in the deck forward of the mast makes a safe seat when doing anchor work and changing staysails, and the forehatch is raised well out of the way of breaking seas.

The sail-plan is simple and cheap; there is no track; the sail runs on hoops in the old-fashioned way. In a small cruiser there is no necessity for the complicated rigging so common in modern practice. The real Bermudan in Bermuda had mast hoops; the Chesapeake Bugeyes used them, and the Dutchman uses parrels on a high mast with a very short curved gaff and often with no shrouds. All that is necessary is to make the mast a little stouter. Several of my designs have been built with hoops and the owners tell me that they work perfectly. I rigged my X-type boat *Moyezerka* with hoops and I was quite satisfied with the effect. If we adopt this plan we must consider how best to deal with the head of the sail when it is reefed. The shrouds will probably not be fitted to the mast-head and there will be a loose foot or so above the hoop, so the head will tend to sag off when the reefs are in. Probably, if the shrouds are well up the mast the head will look after itself, but if they are lower down we must adopt some method to hold it. In *Thuella* I have drawn two mast hoops at the head; if the sail is to be hoisted with a reef in, the extra hoop is attached to the head of the sail, but of course the hoop cannot be attached without lowering the sail. If the shrouds are lower down a jackstay will be necessary. This is led inside the hoops and set up to the deck with a bottle screw, and the top of the sail runs up this jackstay on shackles. But even the jackstay tends to sag away when the sail is reefed. This can be prevented by an arrangement shown in the drawing, which is self-explanatory. There are other methods of achieving the same result. The single-hander's spinnaker can be used with this "mast-hooping" rig; the sail has the shape of an isosceles triangle and its foot is tacked down at each corner to a boom, while its head is run up a jackstay on a shackle. The boom is also fitted with a

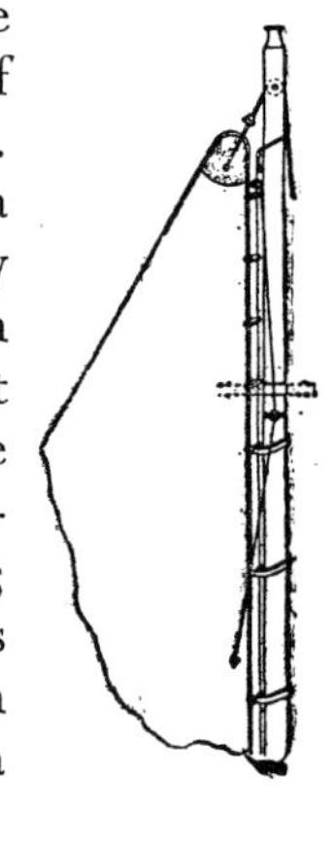

jackstay running its whole length and on this runs a shackle that can be attached to a strop around the mast. There are two guys, one on the starboard, the other on the port side. This spinnaker can be sheeted amidships when it is desired to run without the mainsail, or it is pushed out to one side of the yacht or the other as may be desired. It would be quite easy to arrange two lines from the boom ends to a deadeye on the strop round the mast; then the spinnaker could be swung over without leaving the cockpit. Naturally, the foot of the sail is cocked up, but I do not think that this matters, and of course such a sail is not so effective as a normal shaped spinnaker. A parachute spinnaker would be carried in addition, and it would be wise to have one of heavy canvas with a row of reefs points. This would be used for heavy running.

I think that *Thuella* would face any weather that the crew could stand, and I suggest that there be a hatch a foot wide to cover the after part of the cockpit, or two such hatches if the yacht is being sailed single-handed.

DREAM-OF-ARDEN

My late yacht *Vindilis* was designed many years before the metacentric method was enunciated. Her shelf is not crossed but her analysis is far from perfect. She will go to windward unattended, but with a strong wind on the quarter she takes more weather helm than she should. So when I had mastered the metacentric method I set to work to balance *Vindilis*. The result was *Edith Rose*. She had the same dimensions as *Vindilis*, except that she was 6 in. longer on the L.W.L. This yacht was built and proved to be perfectly balanced, but she was a little tender and had to be put down an inch by inside ballast; also I never quite liked the profile of her canoe stern. So I re-designed her stern with a slightly fuller parabola and put her down another inch. Three of these yachts were built, and I heard good accounts of their behaviour but they were only sailed for one season because of the war. It is interesting to note that these three yachts all trimmed differently. One was correct, another was by the head, and the last was by the stern. I called this design *Rose-of-Arden*. I have recently decided that although the uninterrupted curve from the stem to the heel of the keel gives a very attractive profile, yet the more modern form with a toe has such definite advantages that it should be adopted. The centre of gravity of the ballast keel is brought a few inches lower. The length of straight keel makes it easier to pull the yacht against a wall for scrubbing, and slipping is easier. Finally, the older profile postulates a wood keel that is very wide forward and may cut into a lot of wood. This was very noticeable in my design *Zyklon* that

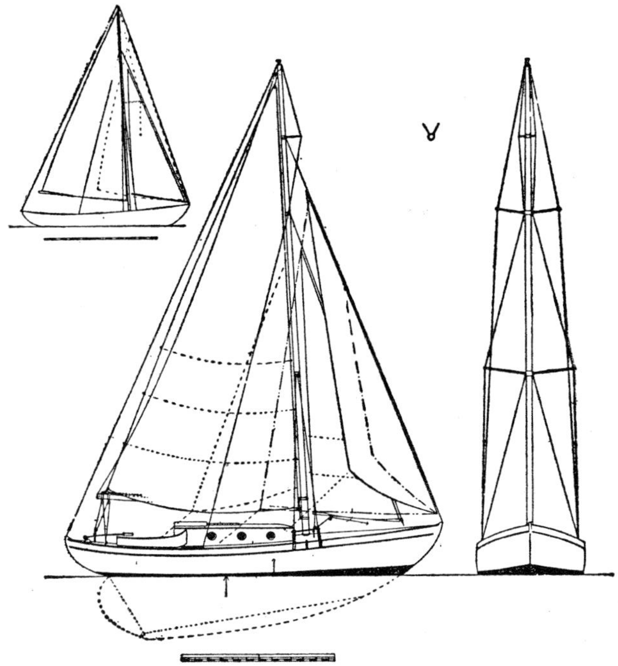

DREAM-OF-ARDEN

L.O.A. 29 ft. 6 in.
L.W.L. 22 ft. 6 in.
Beam 8 ft. 8 in.
Draught 5 ft.

Sail area:
Sloop 360 sq. ft.
Cutter 395 sq. ft.

became the Z 4-tonner. She had an iron keel and I am told that her keel had to be cut from a very wide plank, and this definitely increased her cost. The after deadwood of *Rose-of-Arden* is very thin aft and there is a tendency to weakness. In *Dream-of-Arden* the stern post has been brought forward and the deadwood is shorter and stronger. These alterations to the profile demanded some modification of hull; the sections forward were made fuller and the bilge became harder. These alterations, to my surprise, gave a shelf that is almost straight; it is straight except at the extreme ends where it springs away from the axis symmetrically. The centres of the curves of area coincide so the *Dream-of-Arden* ought to be perfectly balanced.

The cabin and sail plans are those of *Rose-of-Arden* and as they are identical there is no reason for reduplication. The sail-plan shows a sloop and cutter rig. For comfortable family cruising I think that the double headsail is better, for it is so easy to reduce sail by stowing the staysail. The main sheet passes over a large sheave in the cockpit coaming and is sheeted to a cleat inside the cockpit. I found this arrangement excellent in *Vindilis*. One gets a direct pull on the sheet and can look ahead while trimming the sail, often an important advantage.

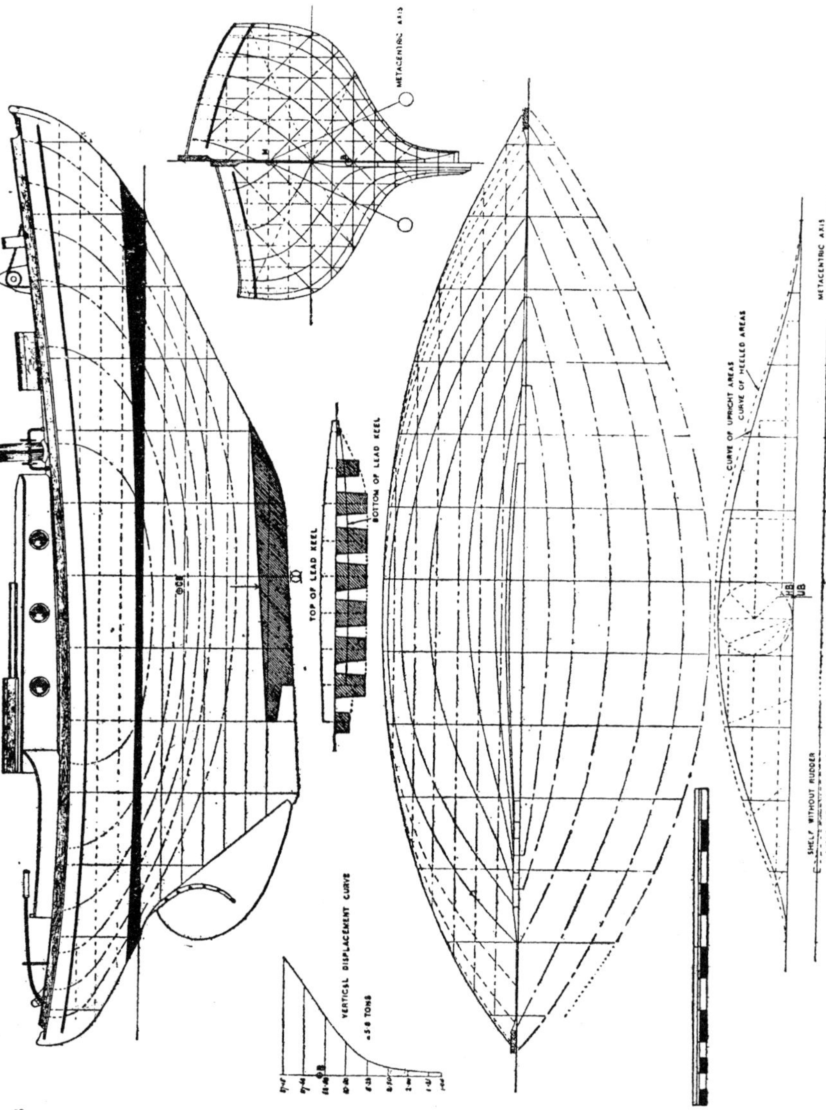

DREAM-OF-ARDEN
Lines

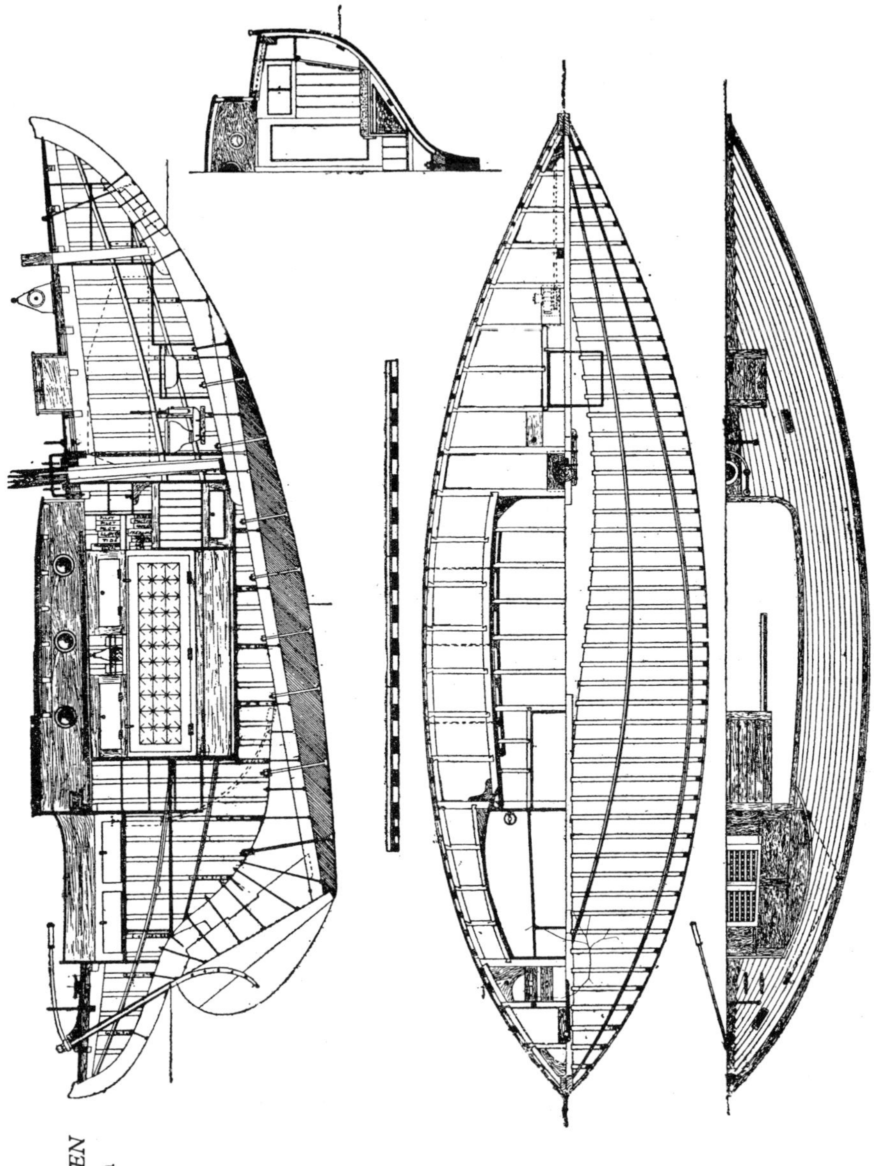

DREAM-OF-ARDEN
Accommodation

The internal lay-out is exactly as I had it in *Vindilis*, except that I have worked in an oilskin locker aft which is most useful. Starting from the forepeak there are two lockers, one for clothes; below this is the chain locker and the coal bunker. Then to port is a mechanical toilet and to starboard is a seat and locker underneath. To port and starboard there are cots. Just forward of the mast is a chain that connects the main beams with the keel and is set up with a bottlescrew; this passes on the strain of the halliards from the main beam to the keel, a precaution that should never be omitted. In the main cabin forward on the port side is a large sideboard that does duty for a chart table, and over it is a locker for navigation instruments. Under the table is a clothes locker opening from above, and aft over the cushion. Behind the cushion is a large locker that holds all the bedding, while above this are lockers. The forward one a clothes locker and the other contains all the cups and saucers. The locker narrows aft and the smaller plates stow aft. Right aft on the port side are food lockers. Care must be taken that if tins are stowed here they are not removed after the compass has been adjusted. There are two magnets to correct the deviation which in *Vindilis* has been reduced to zero. Between the two top lockers on each side is an open locker; one of them houses the wireless set which is a necessity in every cruising yacht. The galley is on the starboard side aft, and aft of it is the oilskin locker. Under each seat is a water tank and the two are connected by a pipe and cock to keep the quantity equal on each side. On each side of the cabin top and over the forecastle door are hand-holds which are as essential below as on deck. The cockpit is fitted with lockers above and below the seats. In *Vindilis* we carried the compass on the port side in the cabin and it was viewed through a window, so we kept our tools and other iron articles in the upper starboard locker. To port we kept food, meat and such like, and the lower lockers under the seats held warps, flares, fog-horn, fog-bell and similar things to port, while on the starboard side were petrol tins, oil cans and engine gear. The small locker under the after seat was used to keep the log line, the watch buoy, winch handles and other articles that might be needed in a hurry. In the larger after locker was the petrol tank, placed fore and aft, reef pendants, small ropes and sail bags. *Vindilis* had thirty-one lockers each with its definite use. The four members of the crew had each a clothes locker. Under the large clothes locker on the port side forward was a large shallow locker to hold the charts in use; others were rolled up in the upper, after, starboard locker in the cabin; a bad plan, but there was no room to keep them all flat. The cabin table folded down under the sole as shown in Fig. 27, page **aa**.

A petrol tank must never be fitted athwartships, otherwise when the yacht heels and petrol is low, air may enter the pipe line and an air lock form. *Vindilis* was fitted with a six horse-power Stuart Turner engine with reducing gear with a central shaft. This was a mistake because reducing gear means a large propeller and therefore a large hole in the rudder, which seriously interferes with the steering and reduces the speed of the yacht. I would prefer to have no reducing gear and an offset installation on the starboard side. A sailing yacht is built to sail, so let the auxiliary be strictly auxiliary. Four knots under power is quite sufficient. Of course, one is really better with no engine, but circumstances often in fact make one inevitable.

KHAMSEEN A

L.O.A.	31 ft. 3 in.
L.W.L.	28 ft.
Beam	10 ft.
Draught	5 ft. 6 in.

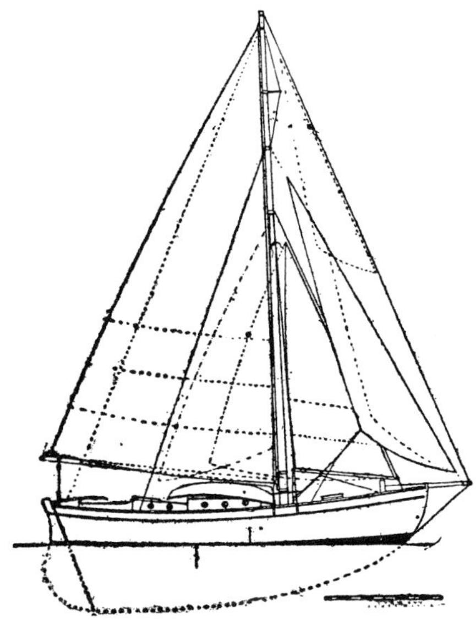

The original *Khamseen* was designed some years ago and two were built; one, *Dorothea*, at Whitstable, and *Mat Ali* in the Malay Peninsula. *Mat Ali* was shipped in a steamer to Port Said and thence she sailed to Cette, and then by canal to Bordeaux and finally to England. Her owner spoke highly of performance, but she was rigged for ocean work, and so she was fitted with a hollow mast and the same sail-plan that I have adopted for *Khamseen A*. Whereas *Khamseen* was planned in pre-metacentric days I re-designed her as *Khamseen A*. The name is the Arabic for "forty," the hot wind that blows in Egypt for forty days. Her lines were descended from *Zyklon*, the Z 4-tonner, and the same sections were used but spaced out to a L.W.L. of 21 ft. Then changing the scale from 1 in. = 1 ft. to

KHAMSEEN A
Lines

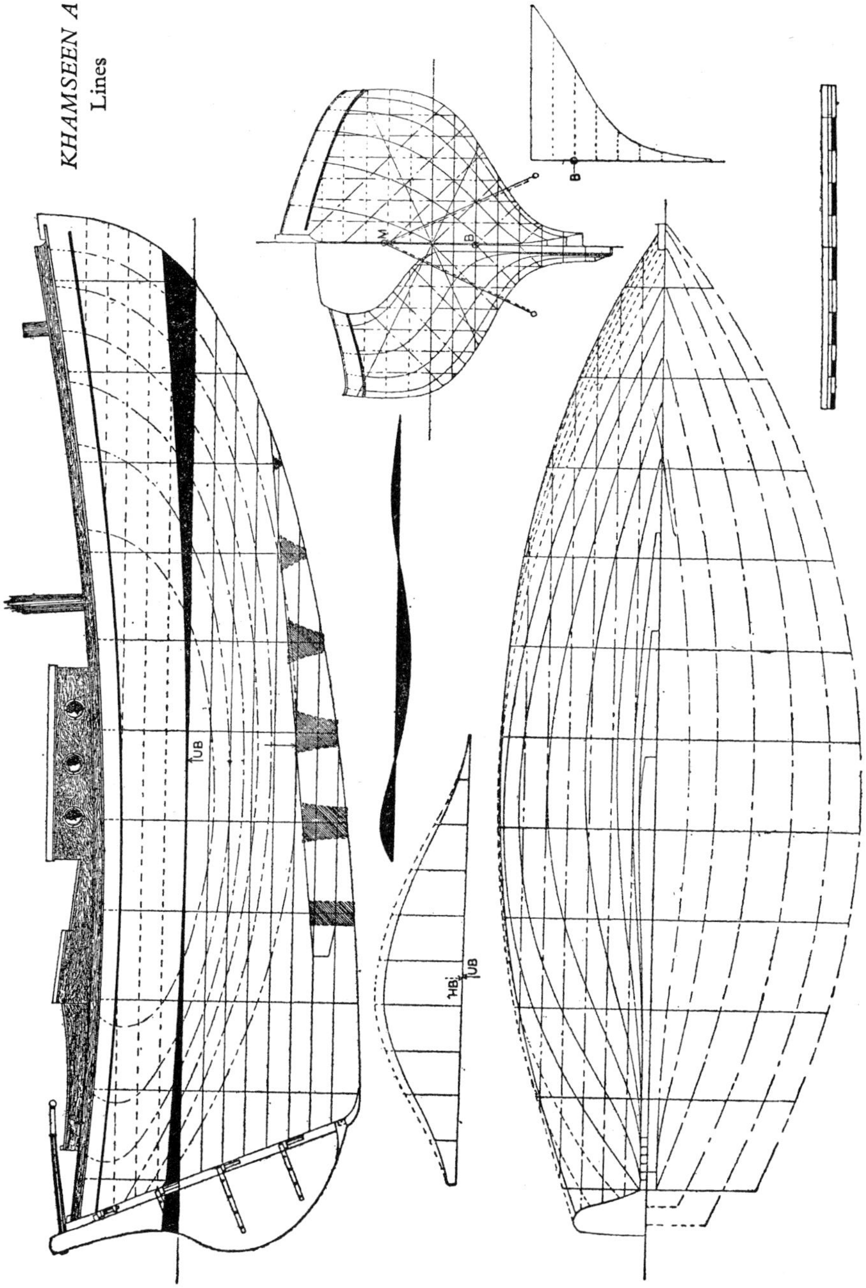

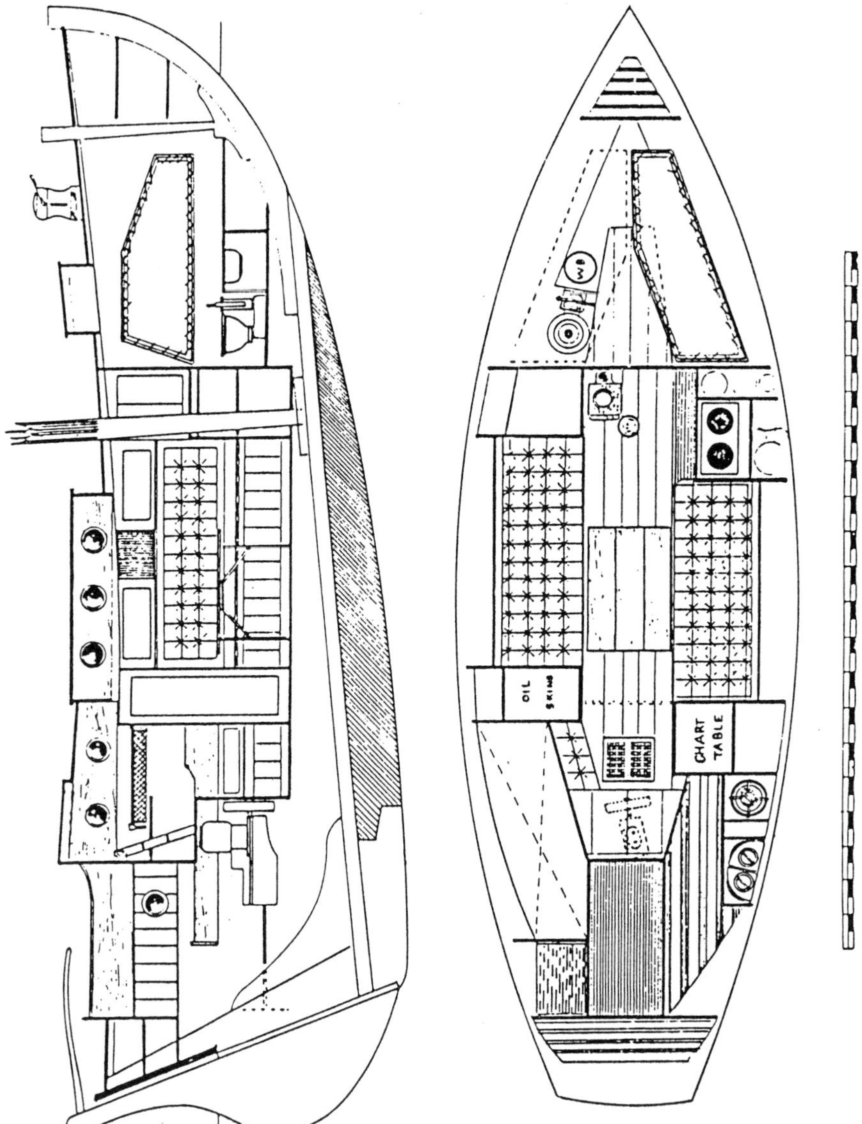

KHAMSEEN A
Accommodation

$\frac{3}{4}$ in. = 1 ft. we arrive at a L.W.L. of 28 ft., and judging by the performance of the Z 4-tonners, the longer *Khamseen* should be even better.

The lay-out would provide comfortable sleeping space for five if a second cot were fitted in the forecastle, otherwise for four. No attempt has been made to fit a second bunk in the after cabin. A galley is shown there, but it could preferably be moved either to the after end of the saloon or to the forecastle (if only one cot were fitted there), thus leaving easily accessible space aft for a chart table and the stowage of sails, warps and boatswain's gear. This latter provision I regard as essential in a real cruiser of this size that is fit to go anywhere.

The sail-plan is quite orthodox. I regard the small jib topsail as a useful sail which is effective out of all proportion to its size.

PRIMA

L.O.A.	22 ft.
L.W.L.	18 ft.
Beam	7 ft. 3 in.
Draught	4 ft.
Sail Area	180 sq. ft.
Sail Area, Genoa ..	114 sq. ft.

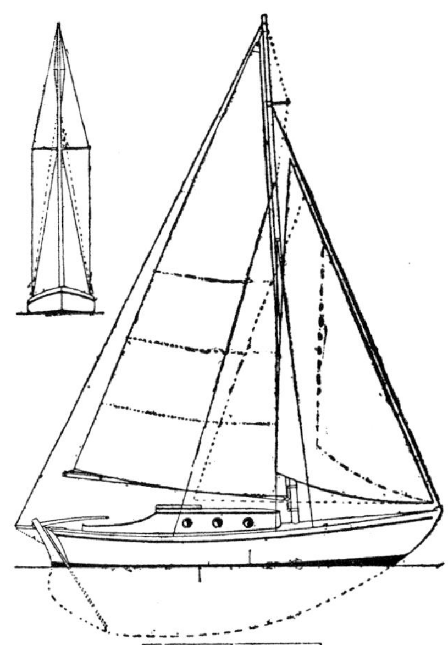

Prima was designed by my daughter, Mrs. R. Jardine Brown, then Miss Ormonde Joan Butler, some seven years ago.* She was entirely responsible for the design and all the hull calculations including the metacentric analysis. Owing to her marriage I was obliged to design the lead keel, the sail plan and the cabin plan. The lead keel is too light and should be at least two hundredweight heavier.

* 1937

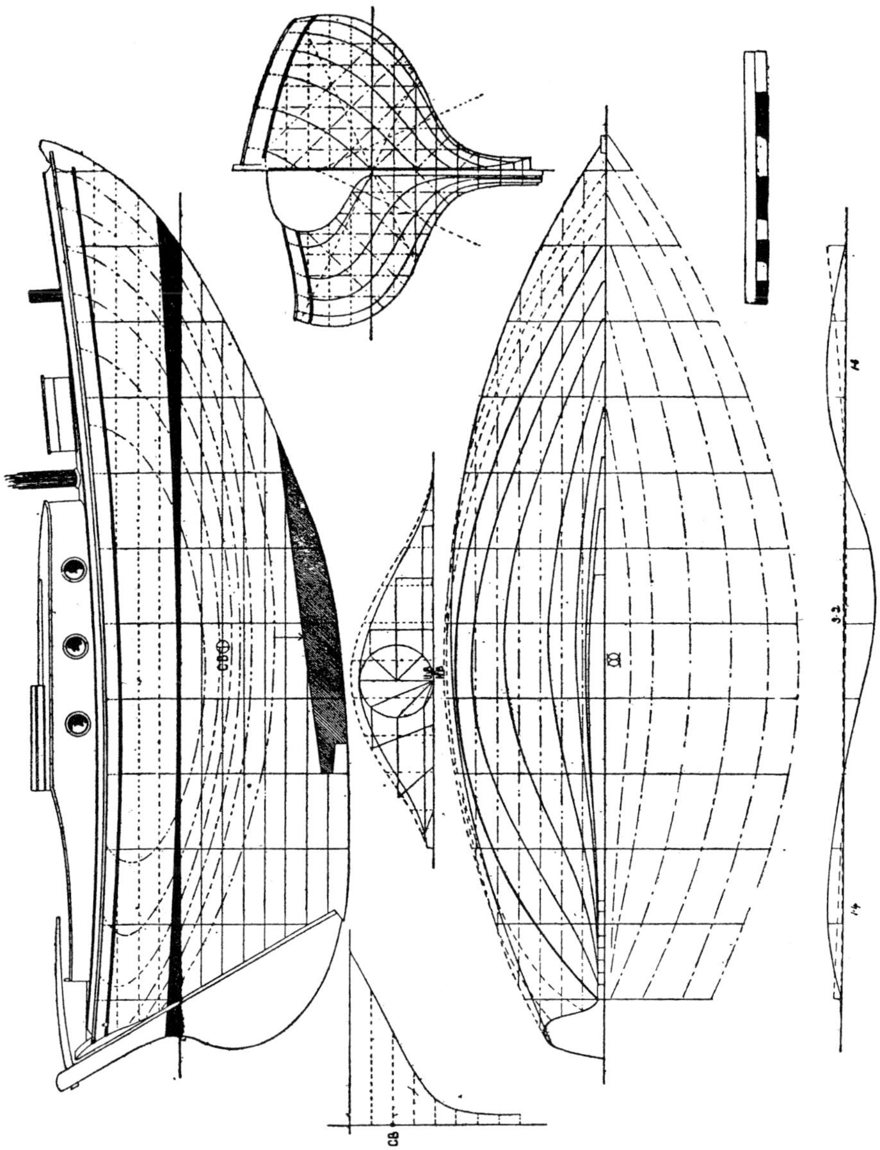

PRIMA
Lines

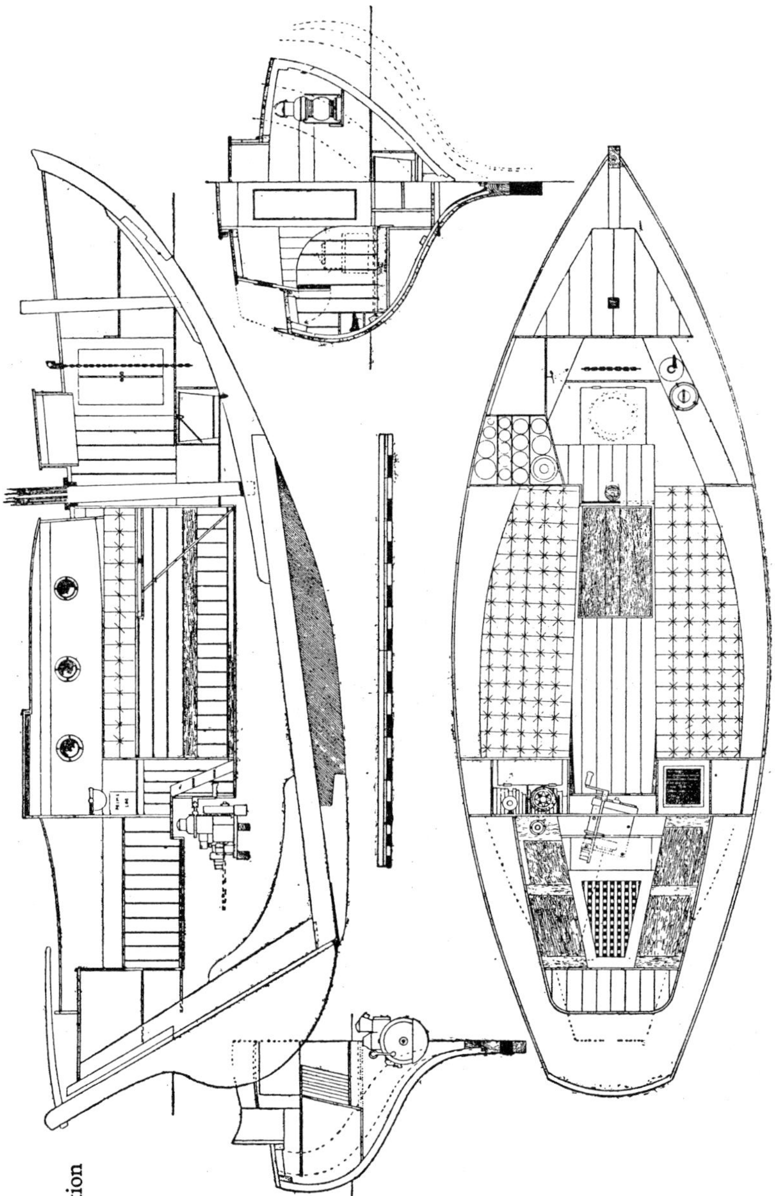

PRIMA
Accommodation

The design appears to me to be an excellent one. She is of Scandinavian design, and if her sections were spaced out to make the water-line 20 ft. and her stern altered to the Scandinavian type as in *Thuella*, she would be an ideal small sea-going cruiser. The tumble-home of the top-sides aft and the rounded transom are original characteristics which give her a personality all her own, but they will not be appreciated by the builder. She should be a fast, handy, seaworthy and comfortable ship fit to go anywhere in any reasonable weather. Her sail-plan is on the small side, suitable for single-handed cruising. The cabin-plan explains itself; there is comfortable accommodation for two. Were I working out the accommodation today I should built up the top-sides amidships as in *Aella* and *Thuella*. This would give a really roomy cabin and a grand deck, and the headroom in the forecastle would be increased. As drawn there will be a little difficulty with the cockpit coaming; the design demands that the sides of the cabin-top shall tumble-home to match the top-sides, and this tumble-home must to some extent be carried aft into the cockpit coaming which will make uncomfortable sitting. This could be met by having a false front to cover the cockpit upper lockers, but it will be far better to have built up topsides. It has been suggested that the built-up deck is unsafe, but I think that one can walk near the middle and I would have a toe-rail six inches inside the outer margin and another in line with the edge of the companion slide; generally the boom would act as a hand-rail. It is no joy to walk along the plank-ways of a small yacht fitted with the ordinary cabin-top. *Prima* is perfectly balanced both metacentrically and by close approximation of the centres of the heeled and upright curves of areas.

FOURTH EDITION: PLANS SUPPLEMENT

All the designs in this supplement (which represents only part of my father's collection) are mentioned in either the text or the biography.

The 22 ft. 6 in. LWL sequence comprises Cyclone II (1919); Yonne (1931); Englyn (1932) and Omega (1936), the metacentric version. Englyn is in the main text so Askadil (1932) and Irmiger (1943) have been inserted. Englyn's sections were spaced out to give Askadil a 25 ft. LWL and Irmiger is the metacentric version of Askadil. They share the accommodation plan.

Cyclone II was derived from the original 19 ft. LWL Cyclone design (1919) which, by somewhat convoluted routes, was the progenitrix of Zyklon (1937), the metacentric version; Khamseen A (1938), which is the enlarged Zyklon, and Maid of Arden, his largest design (1937). Maid of Arden was designed as an ocean cruiser (as were the Khamseens) for a Continental owner who wanted a yacht of a size eligible for ocean racing. I do not know whether she was actually built.

The 23 ft. LWL canoe stern sequence consists of Edith Rose (1936), an improvement on his own yacht, Vindilis (1934) (in the main text); Rose of Arden (1938) and Dream of Arden (1944). Dream of Arden, THB's last design, is in the main text and she and Rose of Arden share the same sail plan and accommodation plan which latter is appropriate also for Edith Rose.

Queen of Arden (1944) is the metacentric version of Aristene (1935). THB left no comments about Sylph of Arden, his penultimate design but she is in the genre he was adopting latterly of canoe sterns and a reverse curve in the forefoot leading sympathetically into a straight keel but with no sharp angle.

As a result of the fifty-two (minimum) Z.4-tonners which were built in addition to the other Zyklons built world-wide, Zyklon has to be my father's most popular design, and deservedly so. Cyclone, in her day, was also a very popular design and several are still in commission.

The plans are all well over fifty years old, which accounts for the variation in the quality of the reproductions.

O.J. J-B, 1994

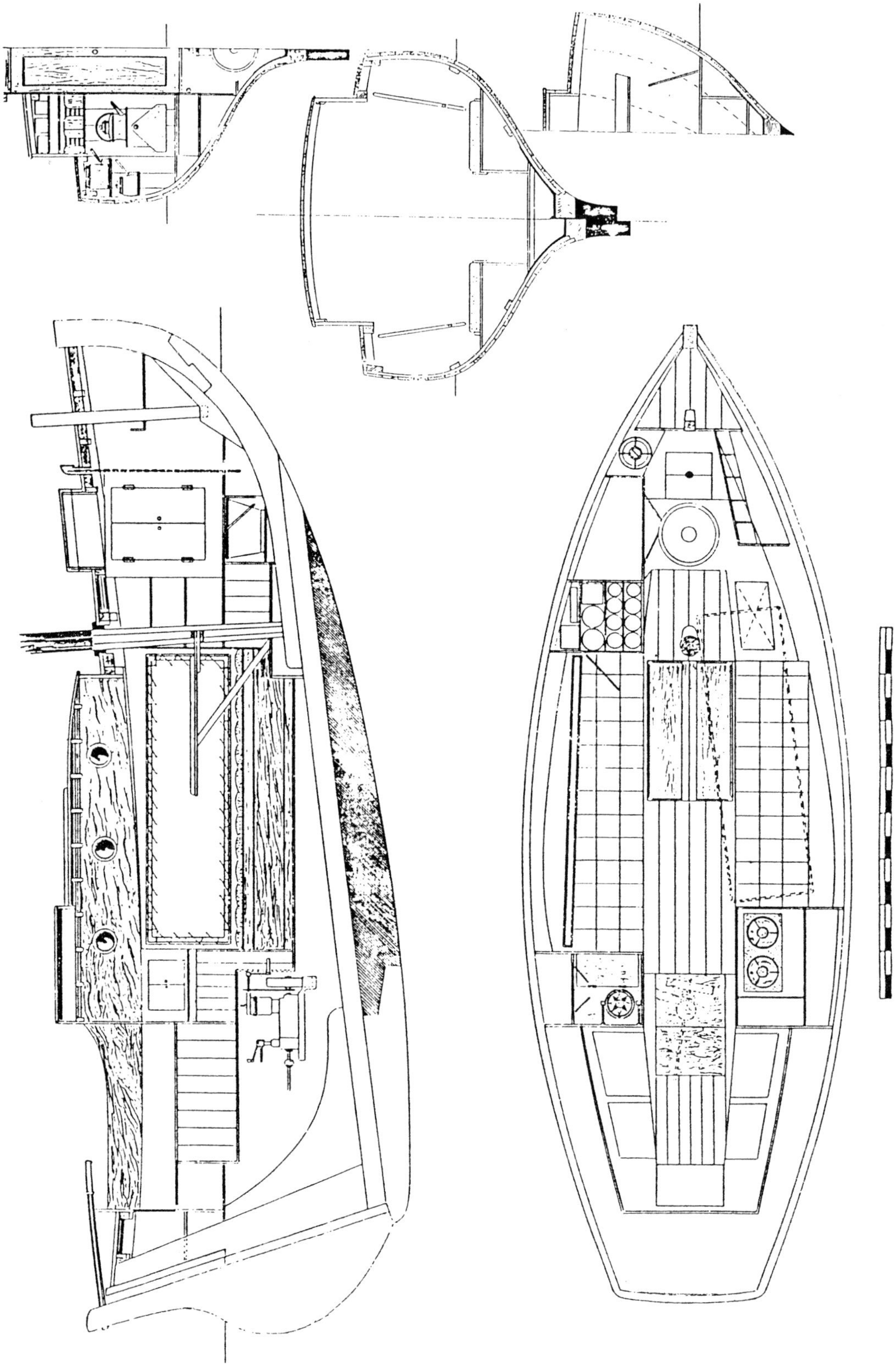

ZYKLON

CYCLONE	
T.M.	3.6 tons
L.O.A.	21 ft.
L.W.L.	19 ft.
Beam	7 ft.
Draught	3 ft. 6 in.
T.S.A.	284 sq. ft.

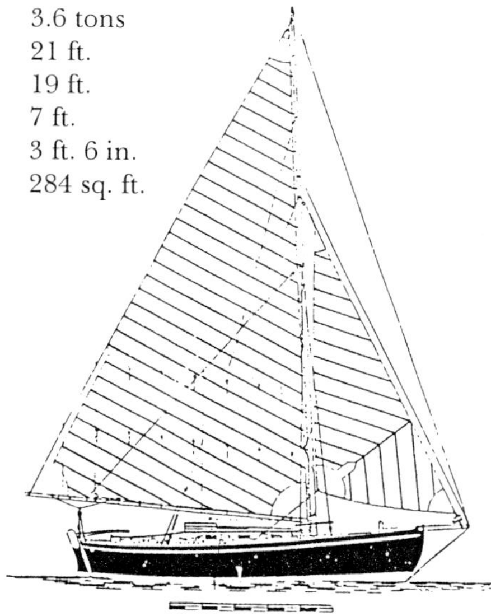

CYCLONE 1926

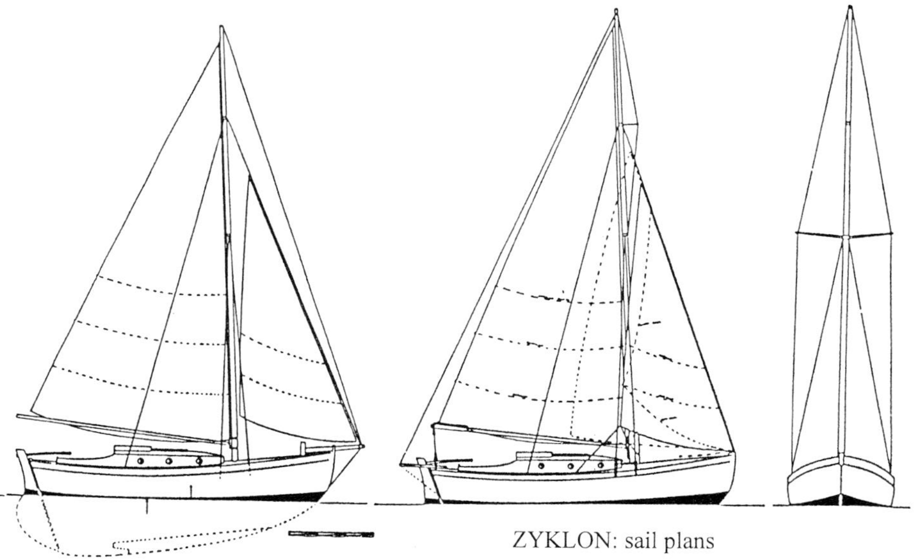

ZYKLON: sail plans

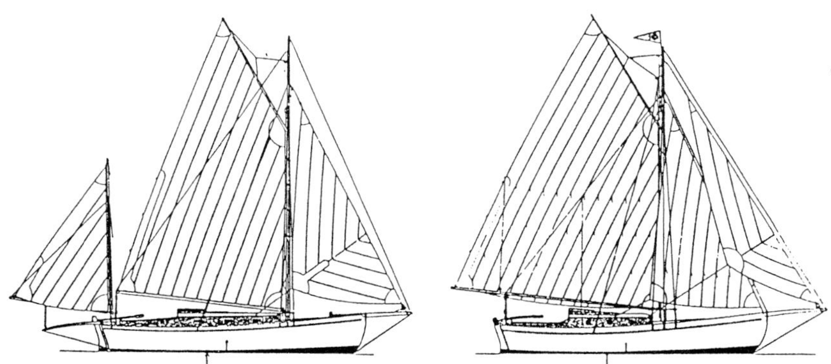

CYCLONE 1919

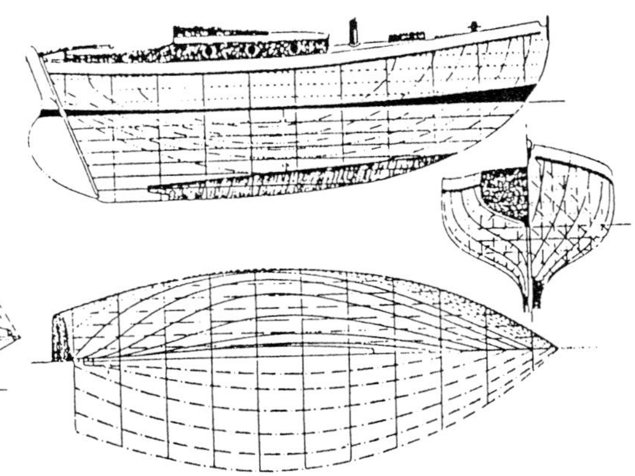

ZYKLON

T.M.	4 tons	Beam	7 ft. 2 in.
L.O.A.	21 ft.	Draught	4 ft.
L.W.L.	19 ft.	T.S.A.	256 sq. ft.

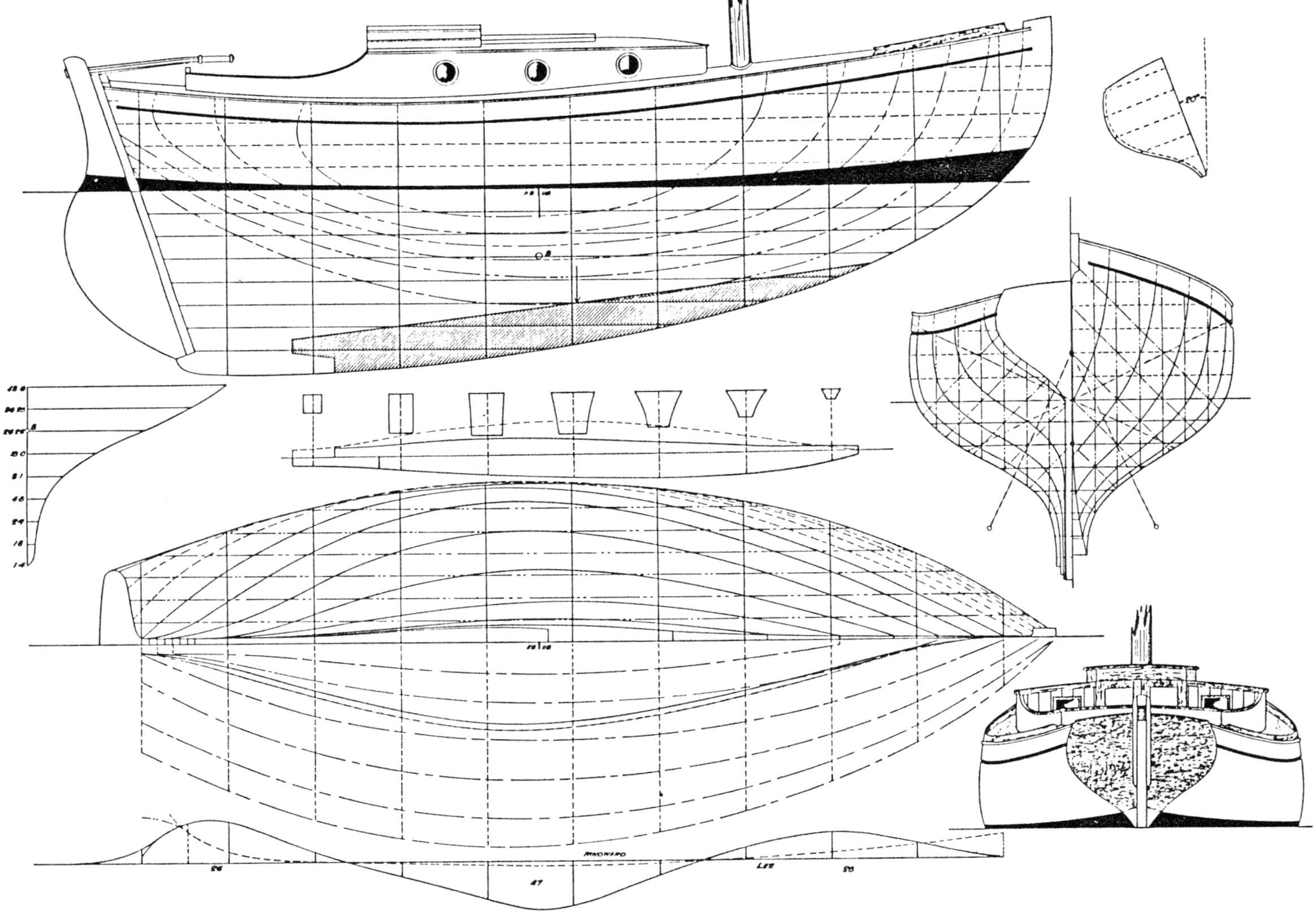

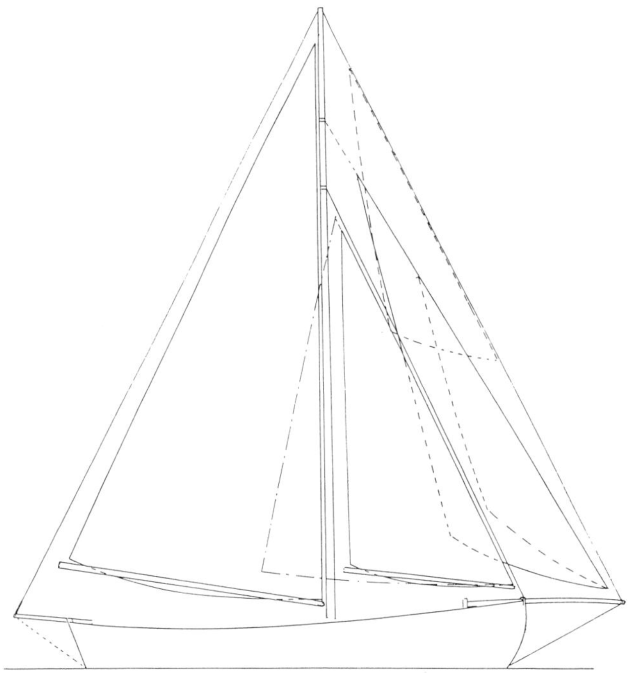

T.M.	6.5 tons
L.O.A.	25 ft.
L.W.L.	22 ft. 6 in.
Beam	8 ft 7 in.
Draught	4ft 2 in.
T.S.A.	360 sq ft

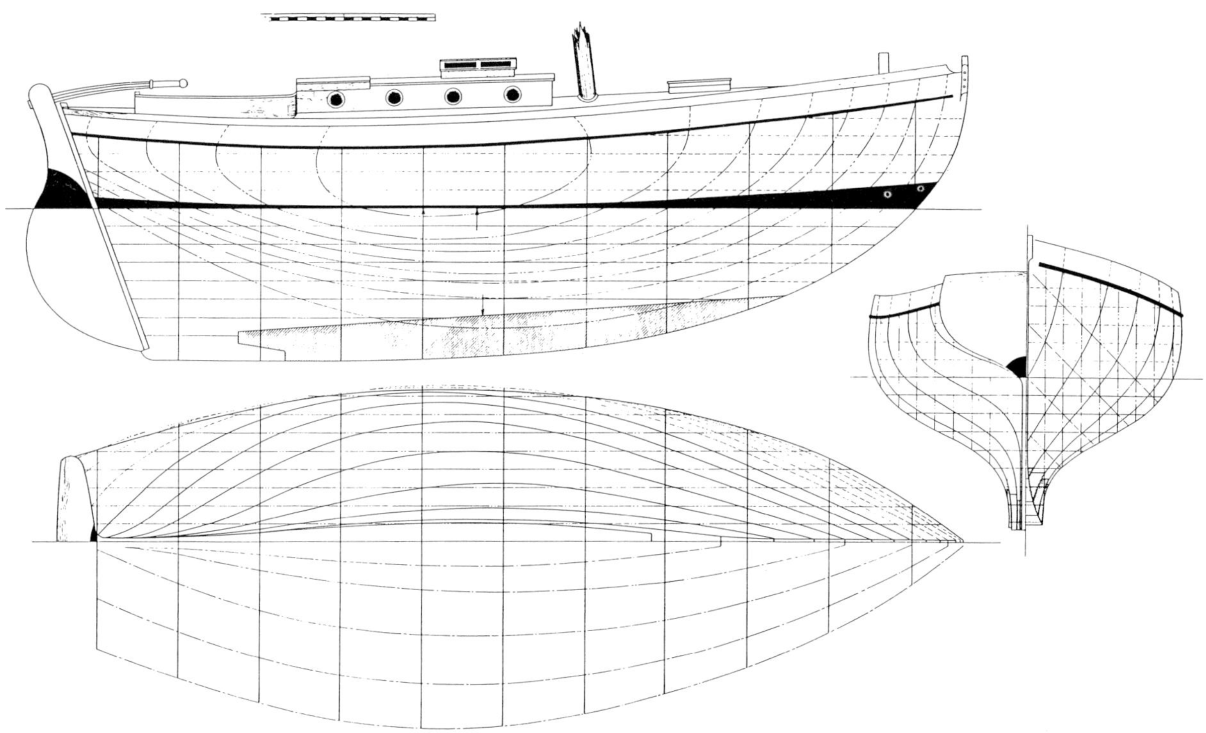

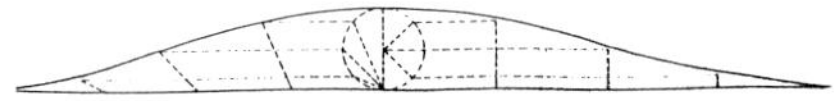

CYCLONE II

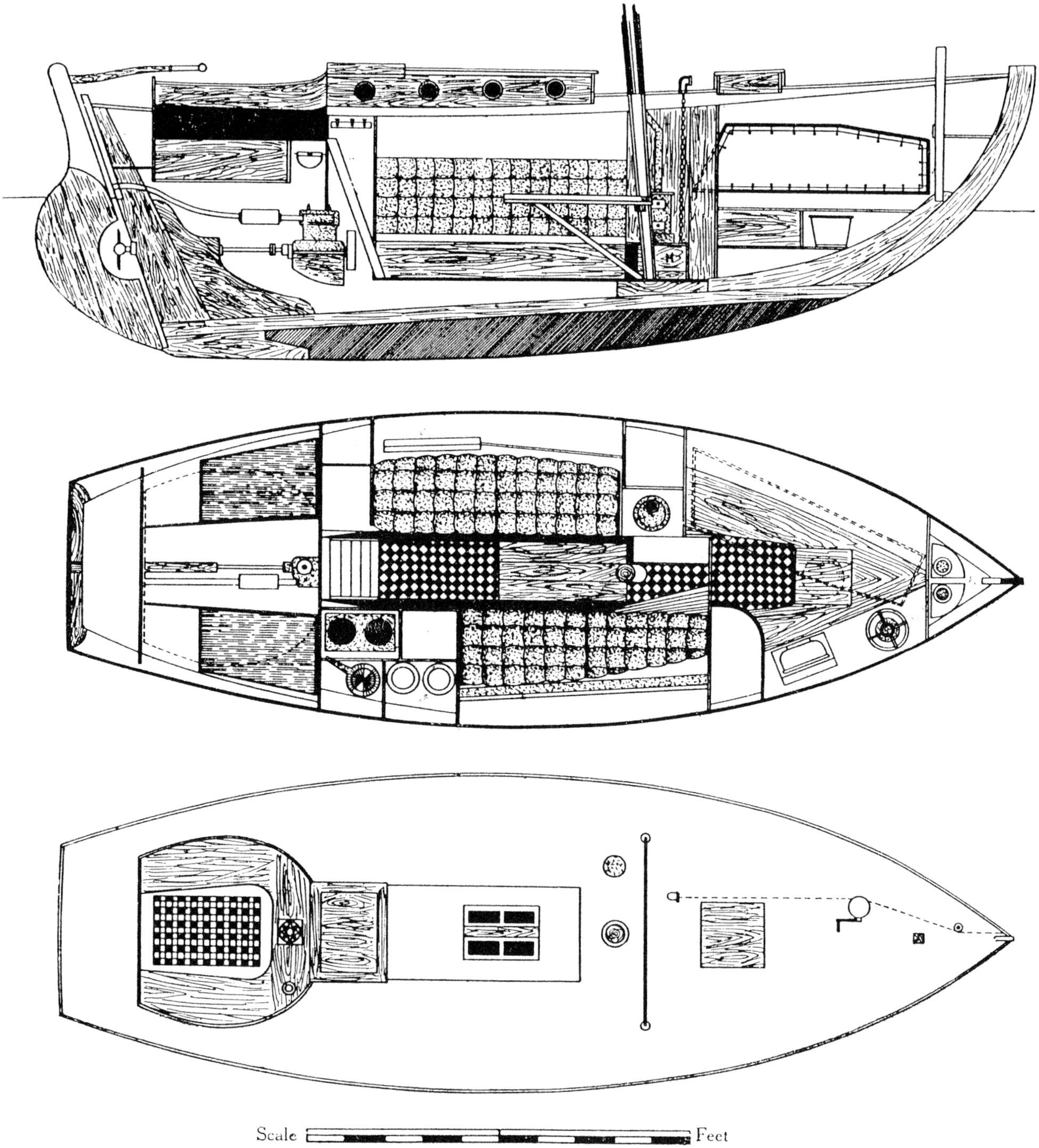

CYCLONE II

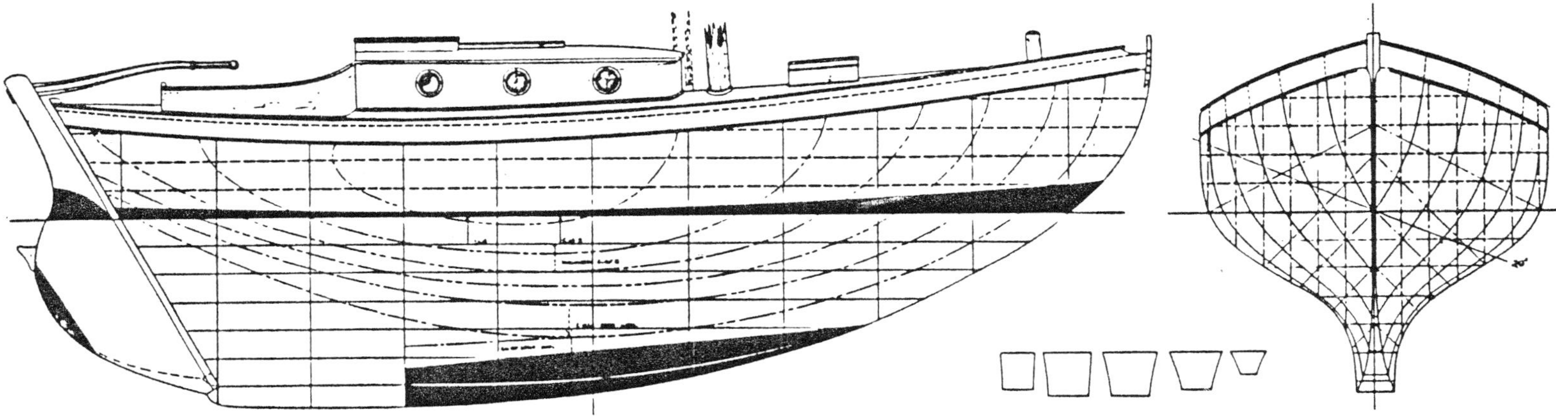

T.M.	6.8 tons
L.O.A.	26 ft.
L.W.L.	22 ft. 6 in.
Beam	8 ft. 7in.
Draught	4 ft. 6 in. or 4 ft.
T.S.A.	393 sq. ft.

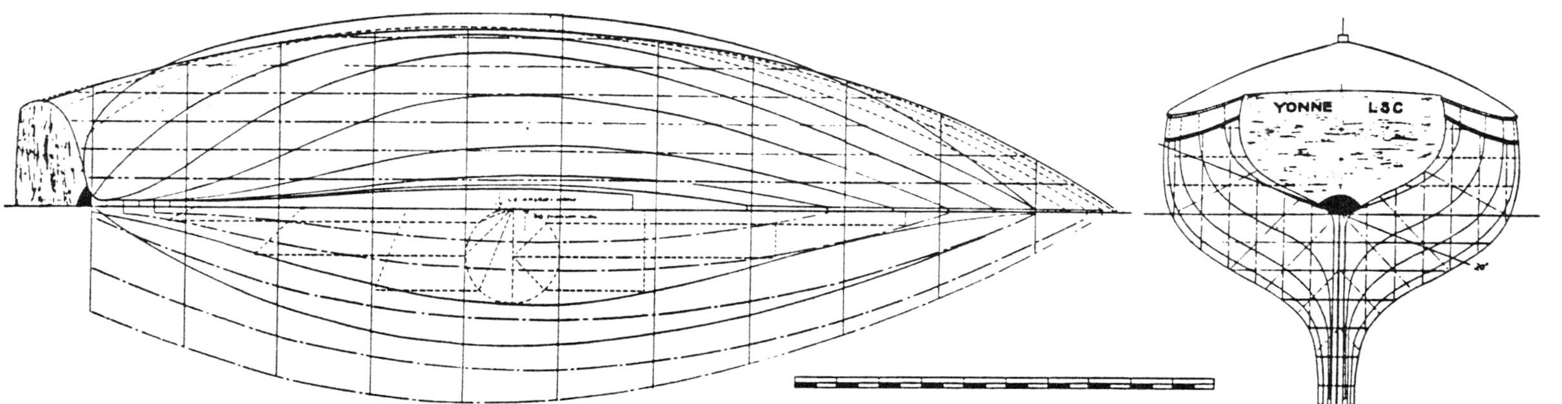

YONNE

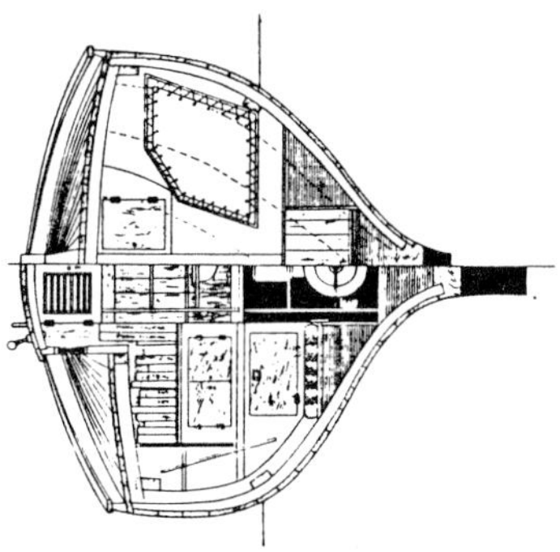

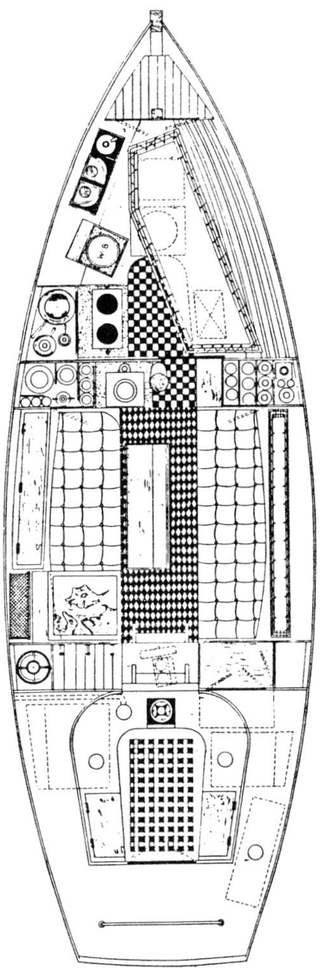

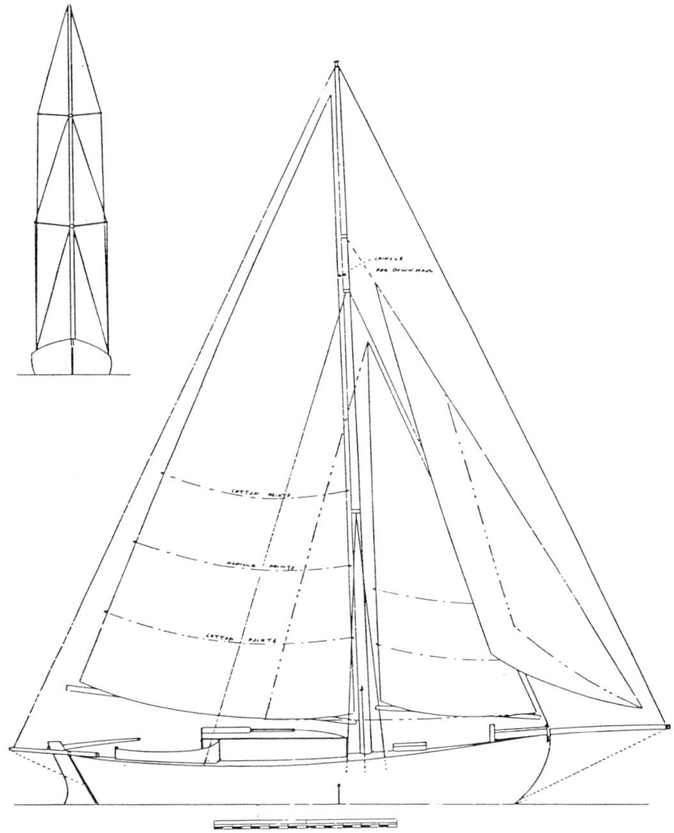

ENGLYN
Revised sail plan Yonne and Englyn

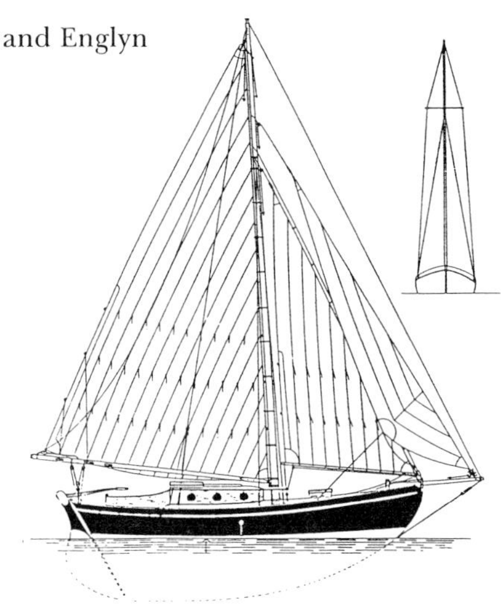

Original
sail plan

YONNE

YONNE: Accommodation and Sail Plans

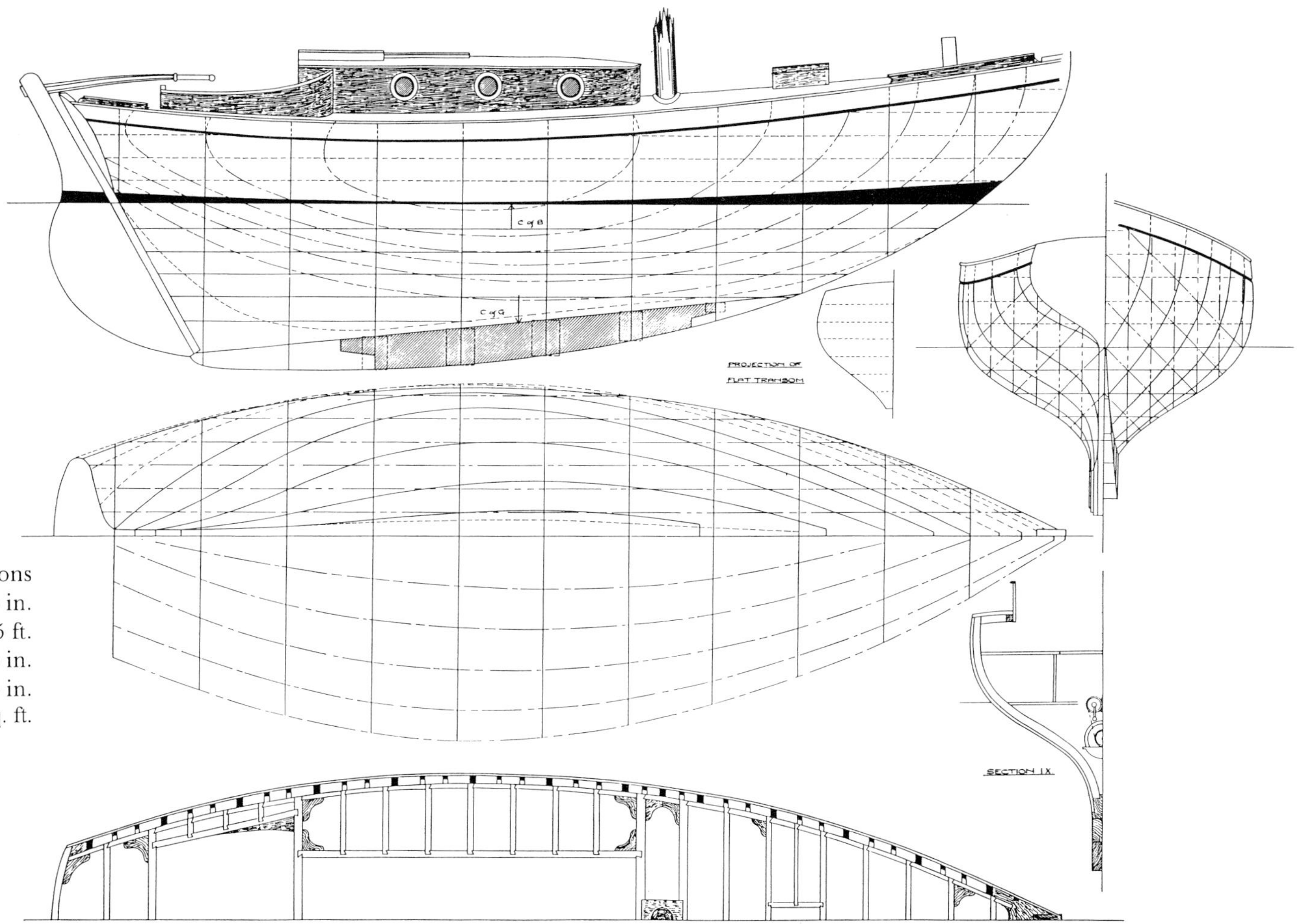

T.M.	8.6 tons
L.O.A.	29 ft. 6 in.
L.W.L.	25 ft.
Beam	8 ft. 6 in.
Draught	8 ft. 6 in.
T.S.A.	514 sq. ft.

ASKADIL

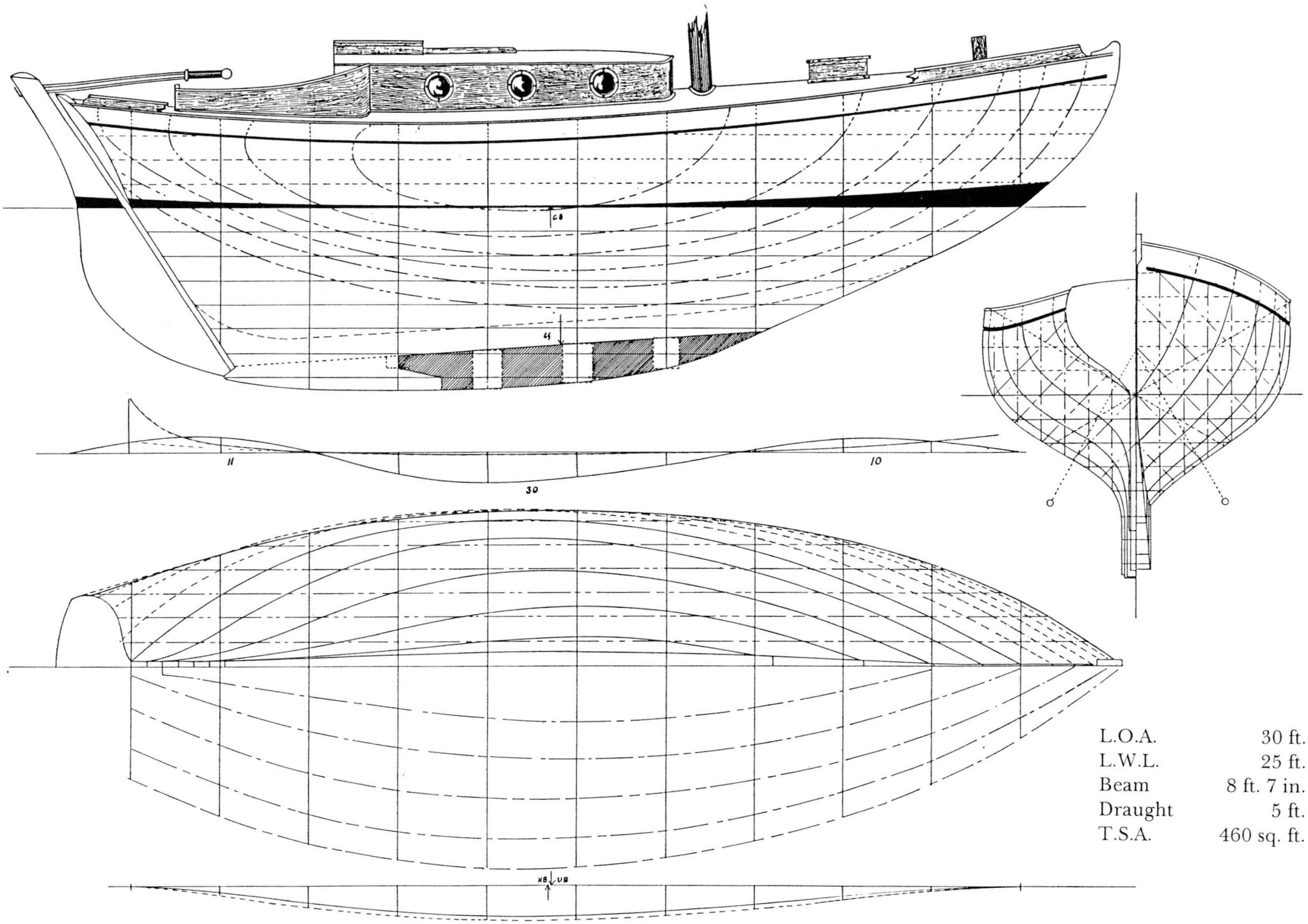

L.O.A.	30 ft.
L.W.L.	25 ft.
Beam	8 ft. 7 in.
Draught	5 ft.
T.S.A.	460 sq. ft.

IRMIGER
The metacentric version of Askadil

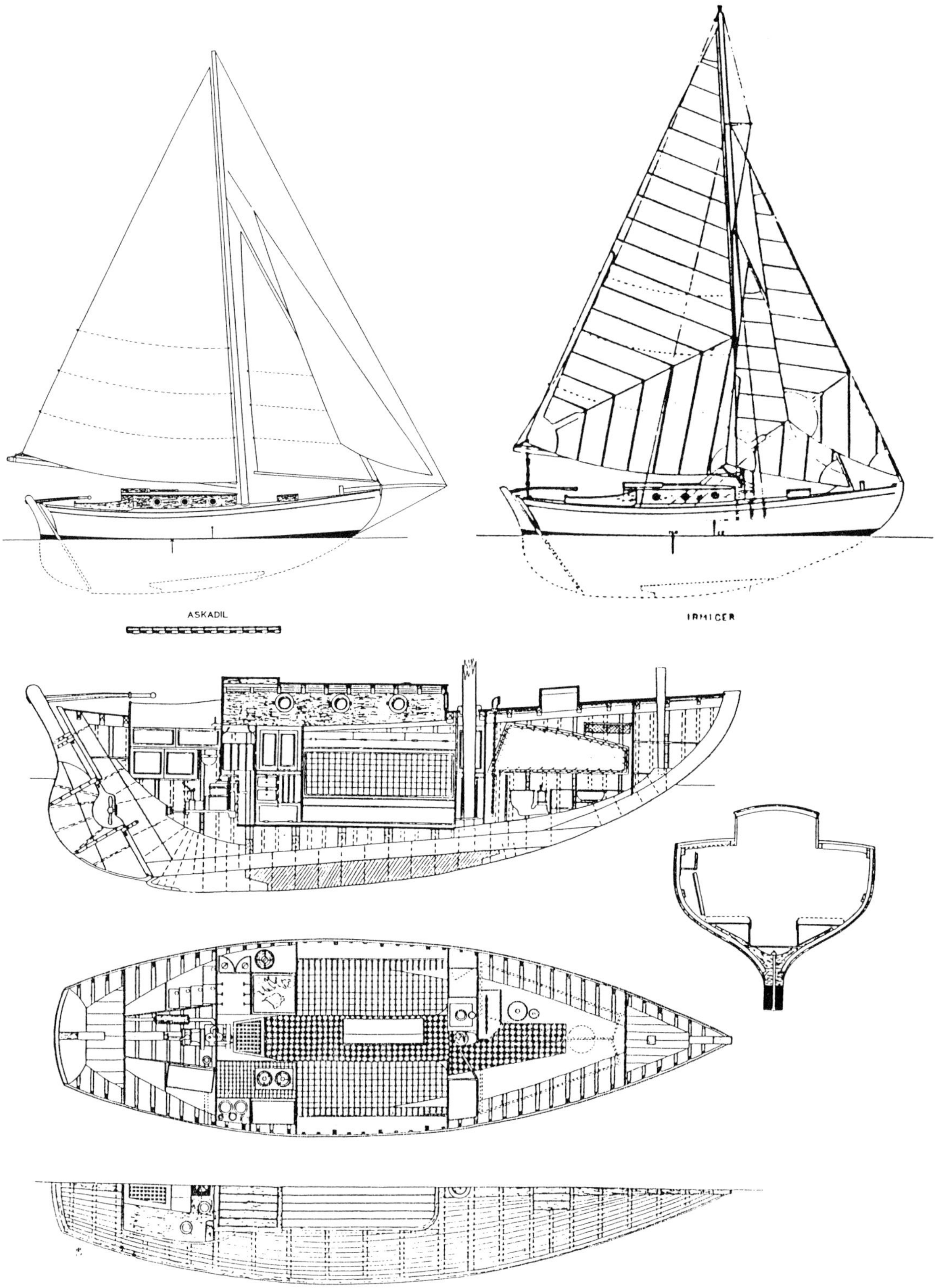

ASKADIL and IRMIGER

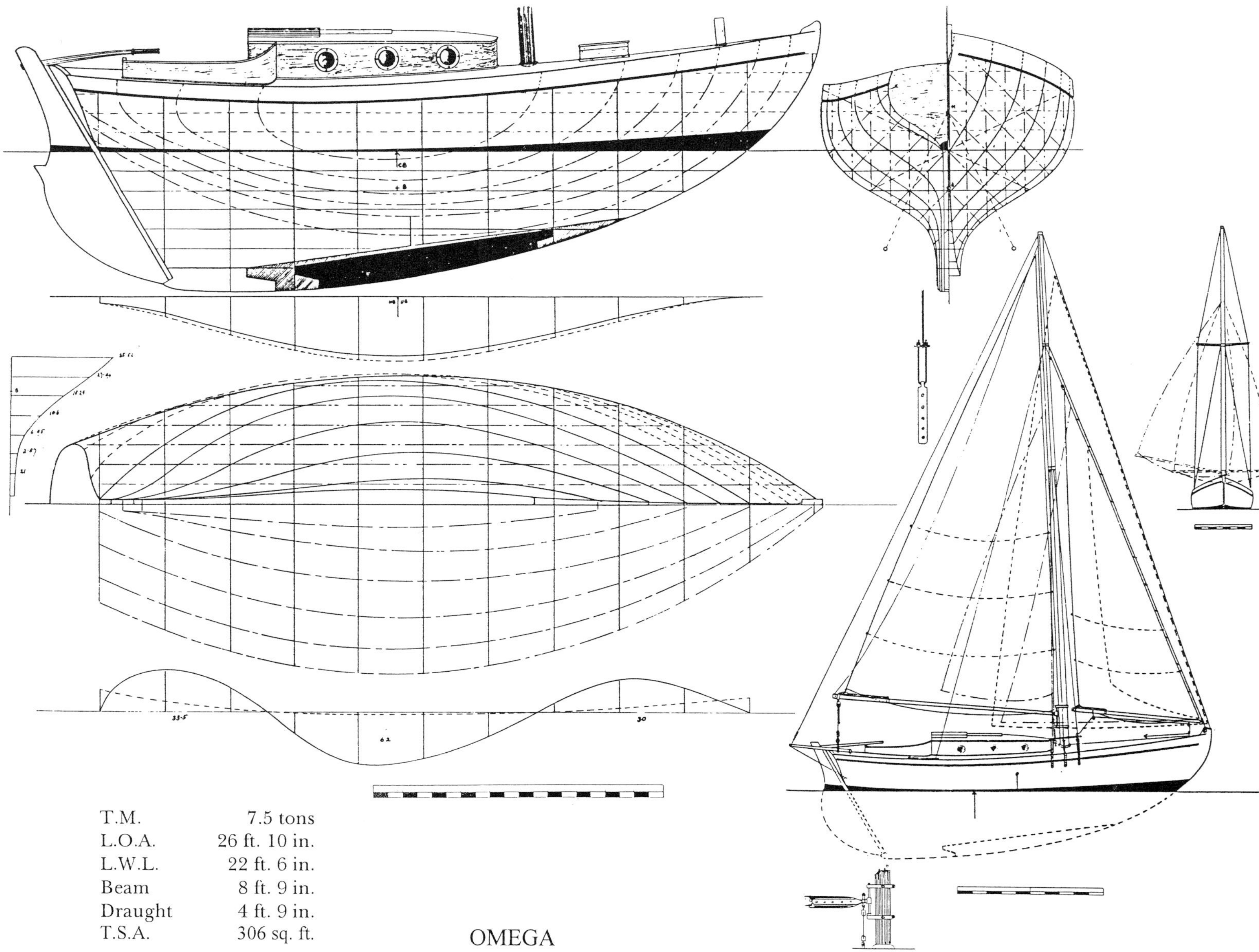

T.M.	7.5 tons
L.O.A.	26 ft. 10 in.
L.W.L.	22 ft. 6 in.
Beam	8 ft. 9 in.
Draught	4 ft. 9 in.
T.S.A.	306 sq. ft.

OMEGA

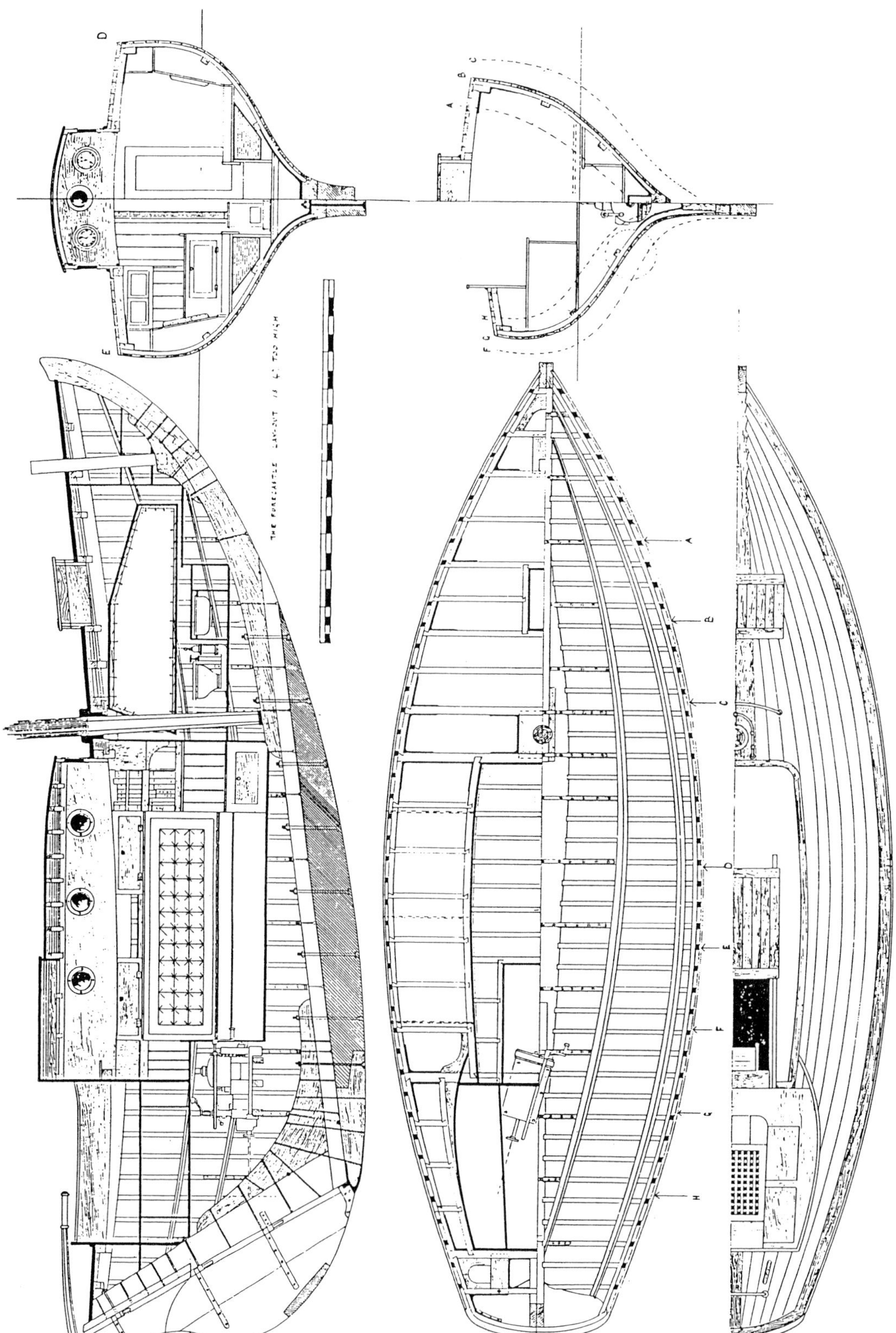

OMEGA: Construction plan

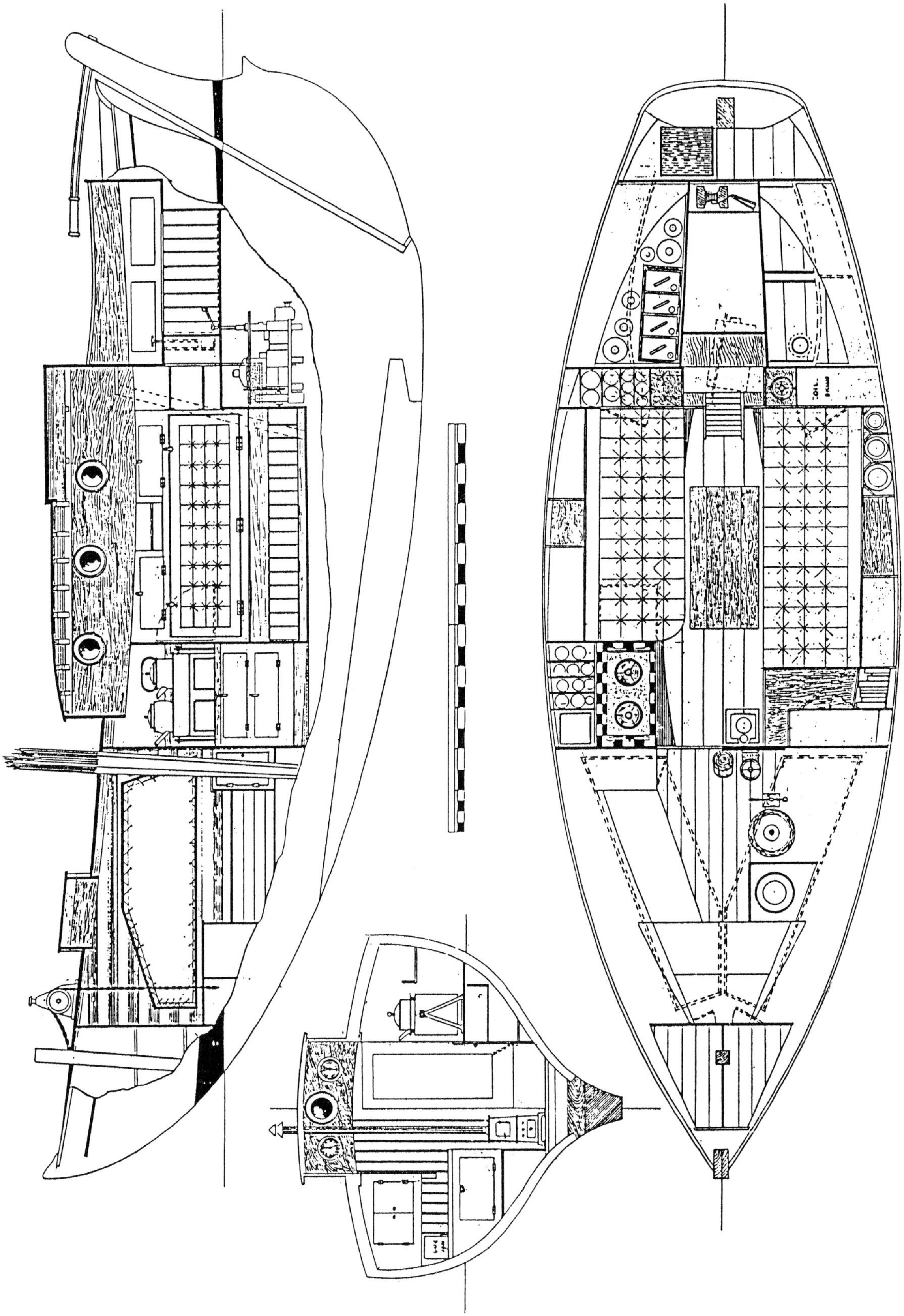

OMEGA: Accommodation

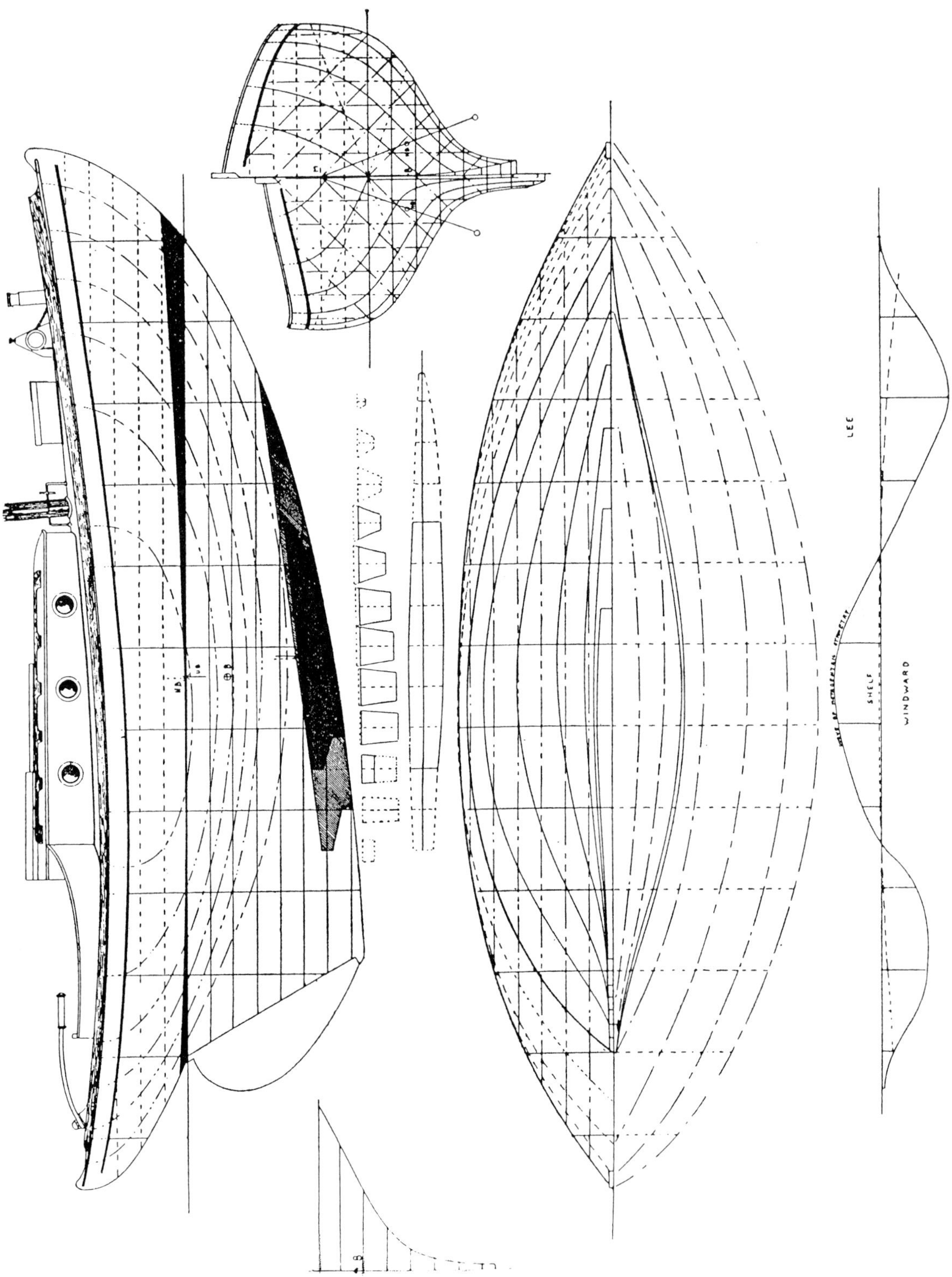

EDITH ROSE, "The improved Vindilis"

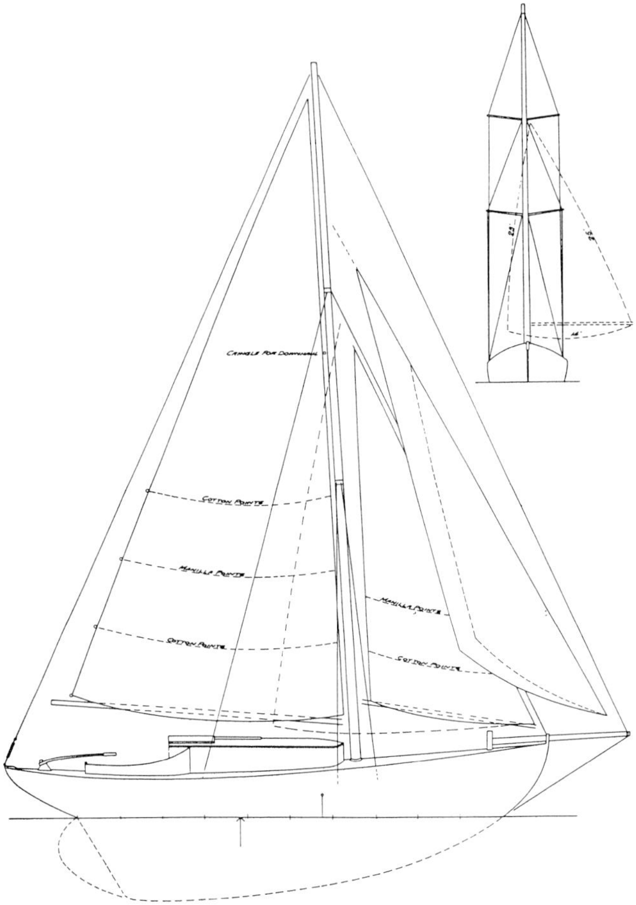

T.M.	7 tons
L.O.A.	29 ft. 6 in.
L.W.L.	23 ft.
Beam	8 ft. 6 in.
Draught	4 ft. 8 in. or 5 ft.

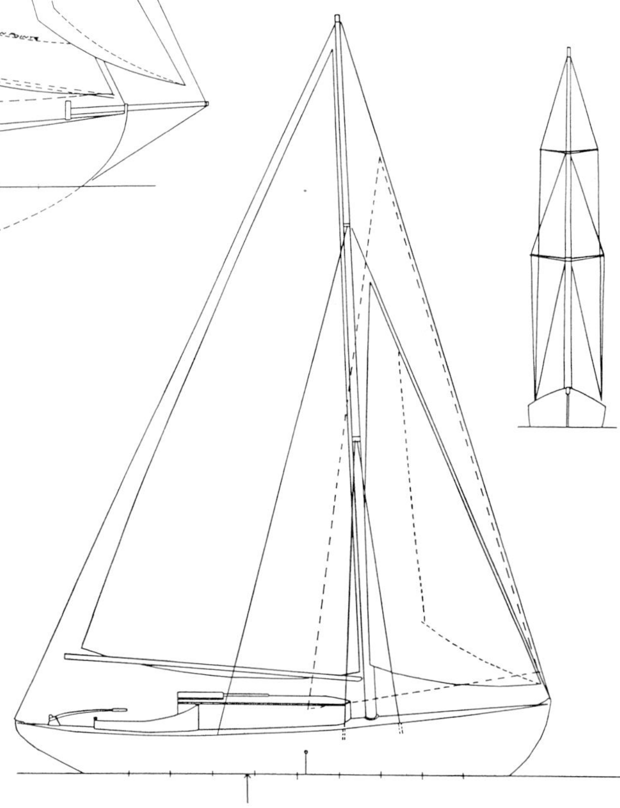

EDITH ROSE

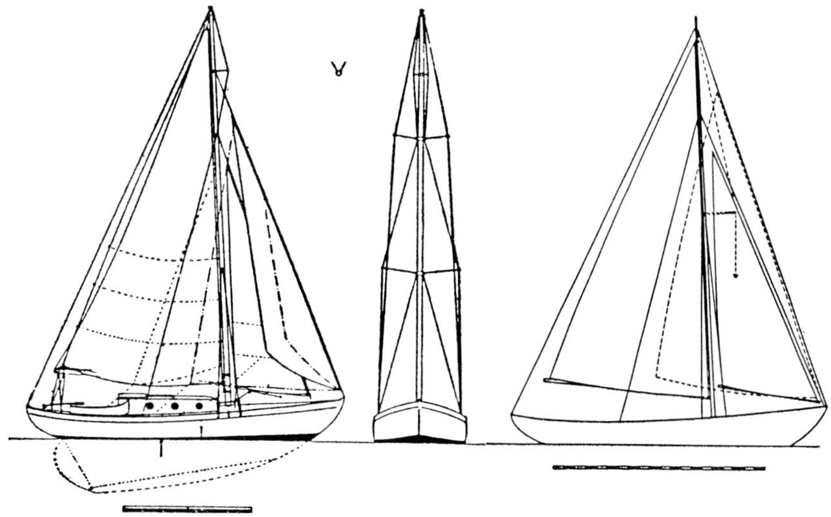

T.M.	7.8 tons
L.O.A.	30 ft.
L.W.L.	23 ft.
Beam	8 ft. 9 in.
Draught	5 ft.
T.S.A.	
Cutter	395 sq. ft.
Sloop	360 sq. ft.

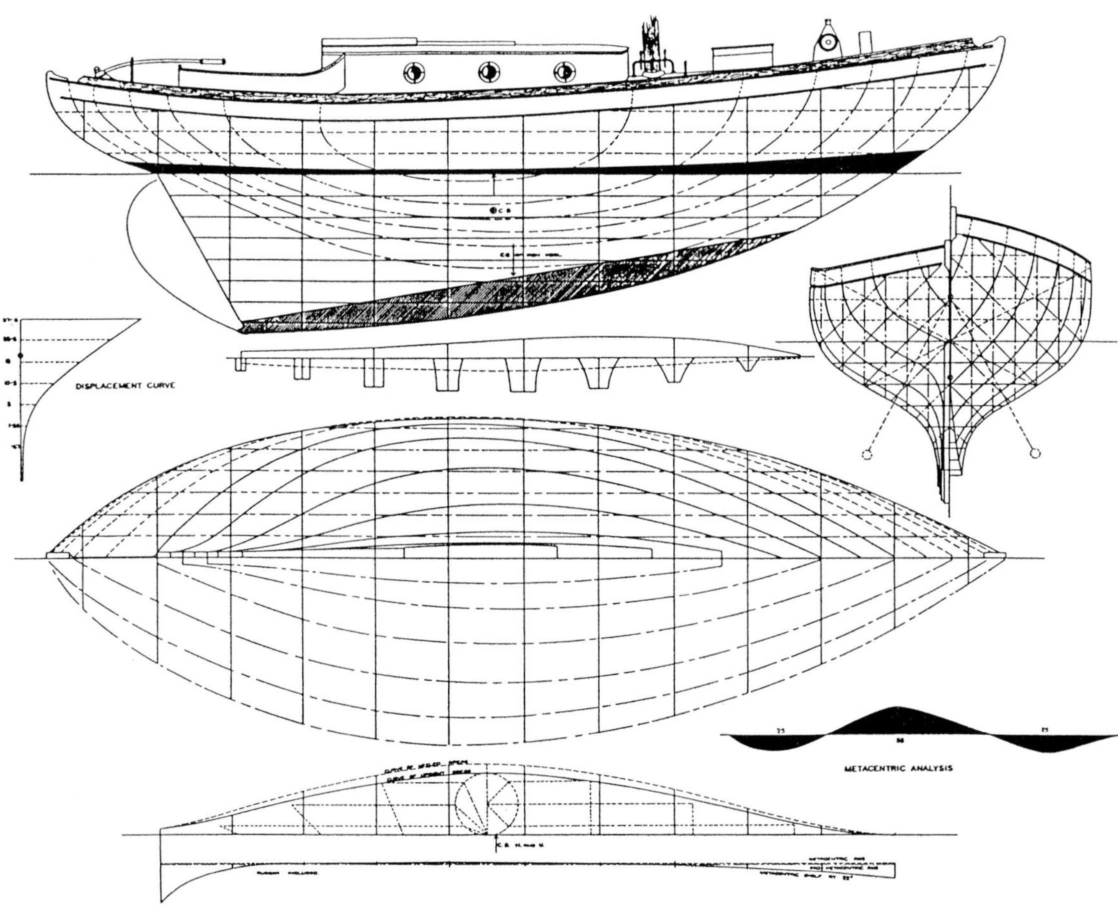

ROSE OF ARDEN

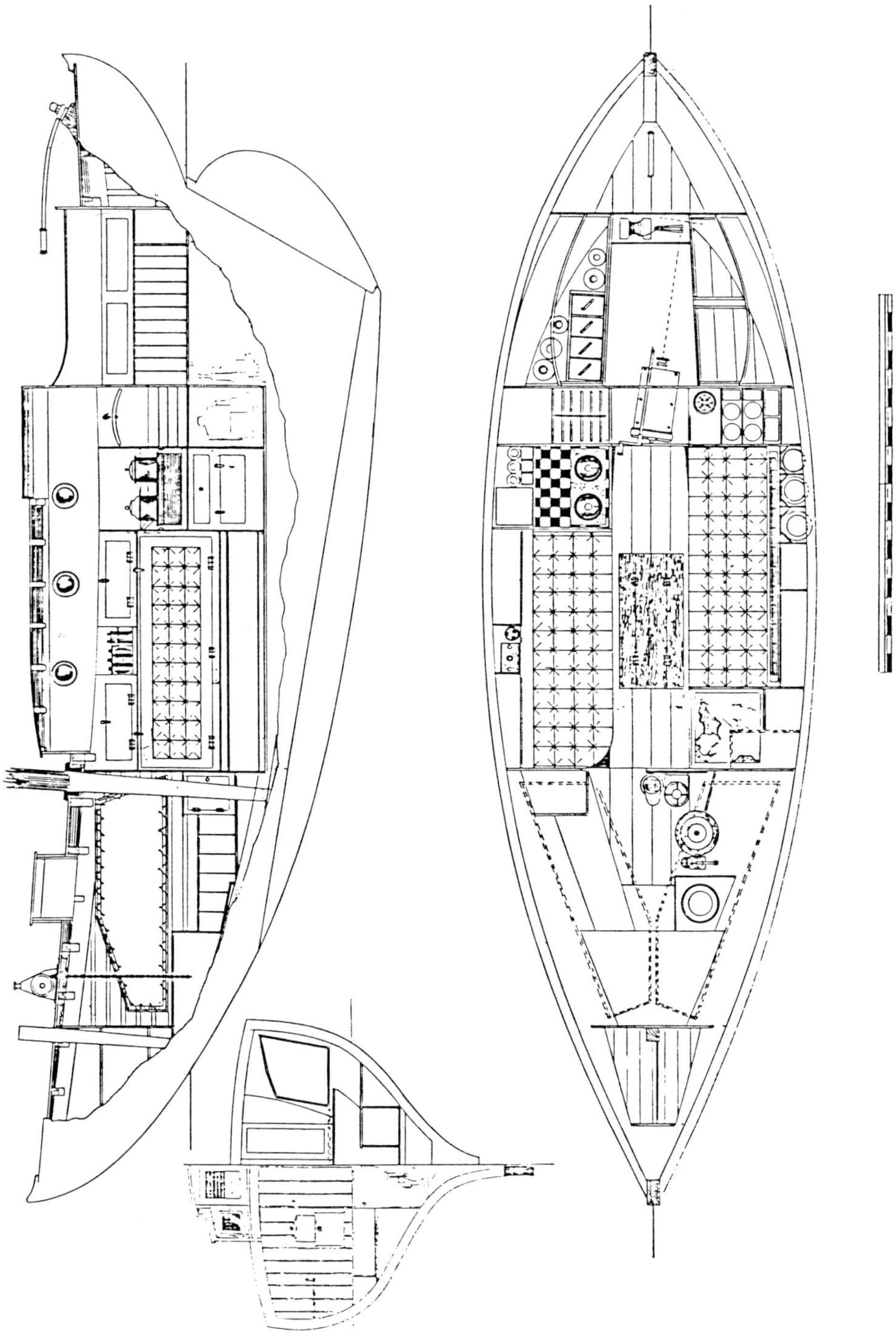

ROSE OF ARDEN: Accommodation

T.M.	19 tons
L.O.A.	44 ft.
L.W.L.	35 ft.
Beam	11 ft.
Draught	7 ft.
T.S.A.	945 sq. ft.

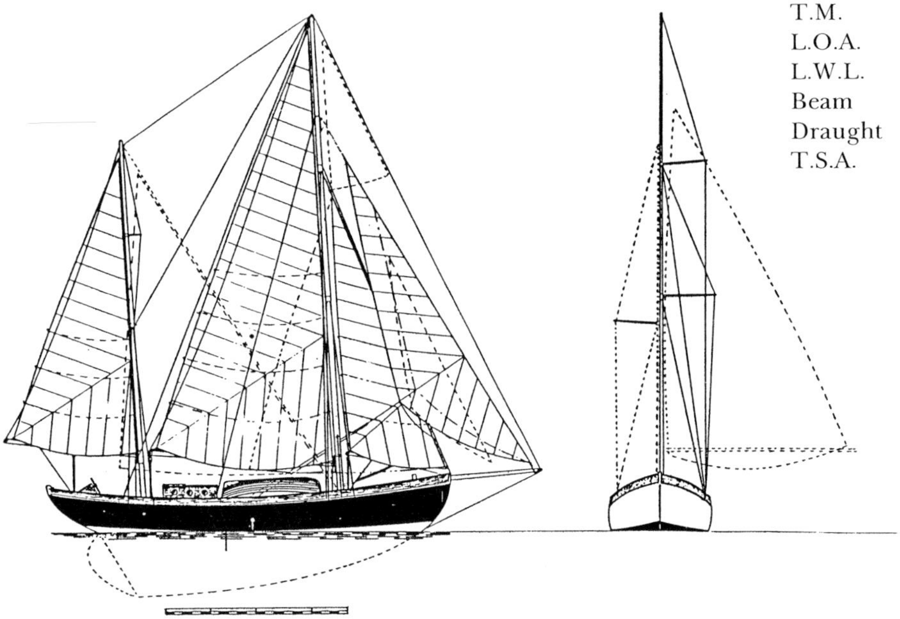

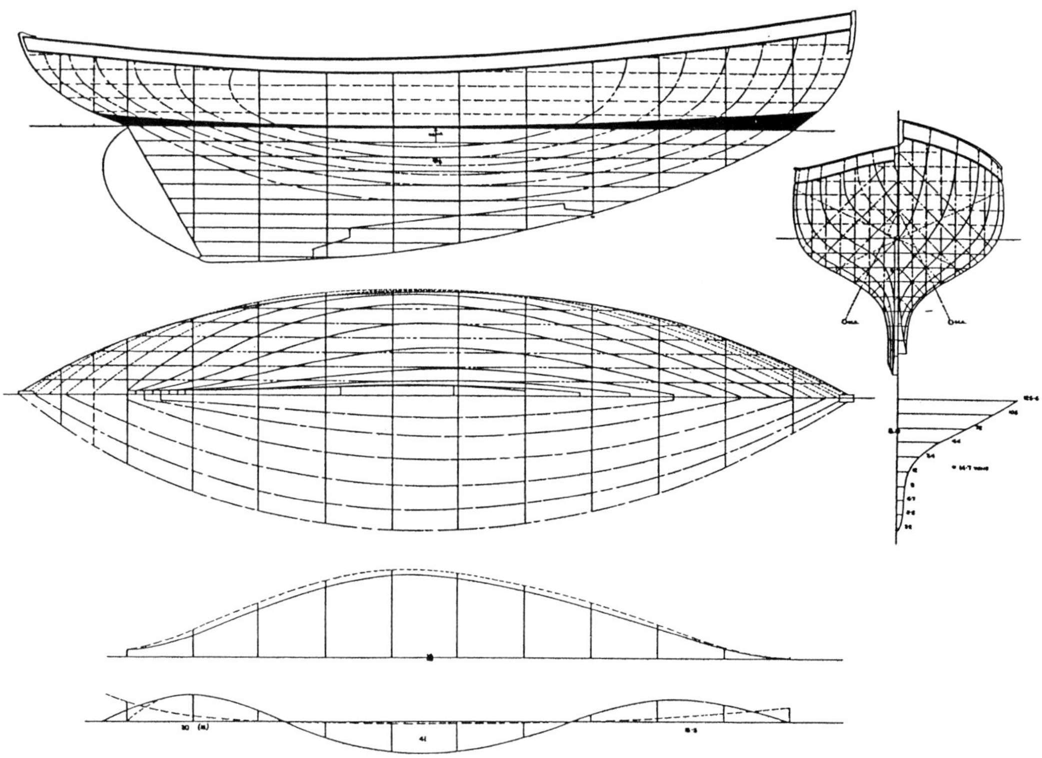

MAID OF ARDEN - THB's largest design

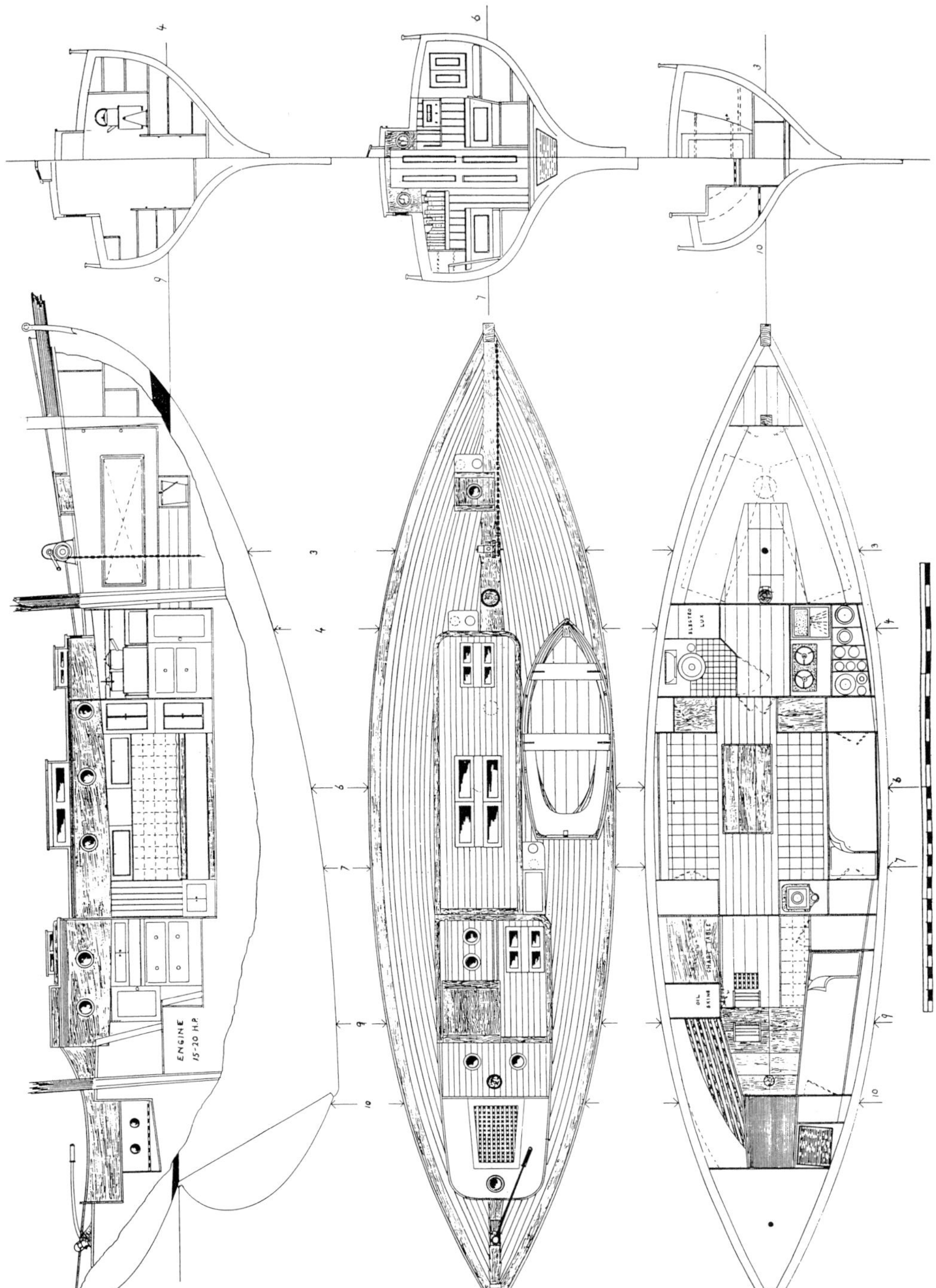

MAID OF ARDEN: Accommodation plan

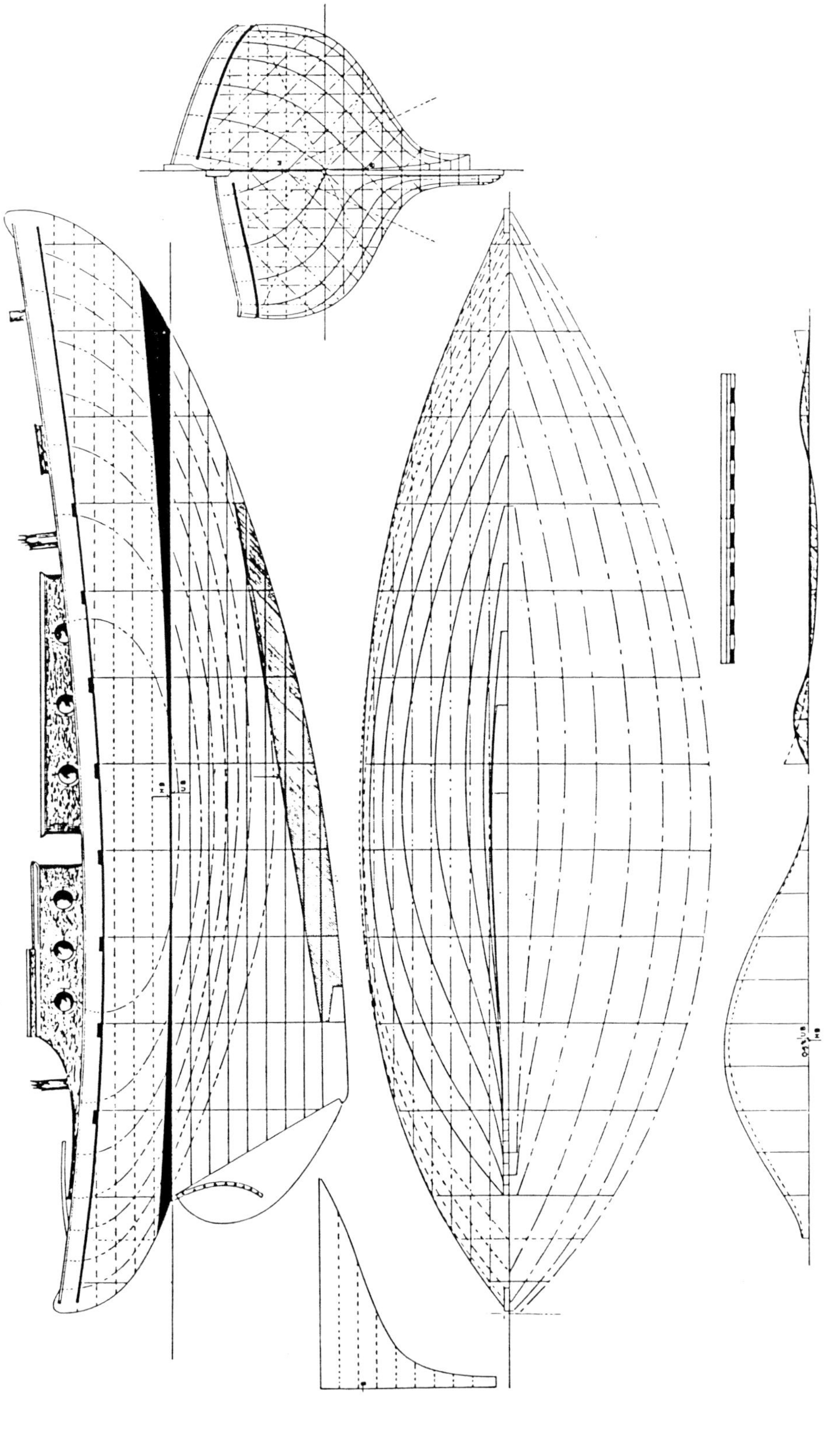

QUEEN OF ARDEN

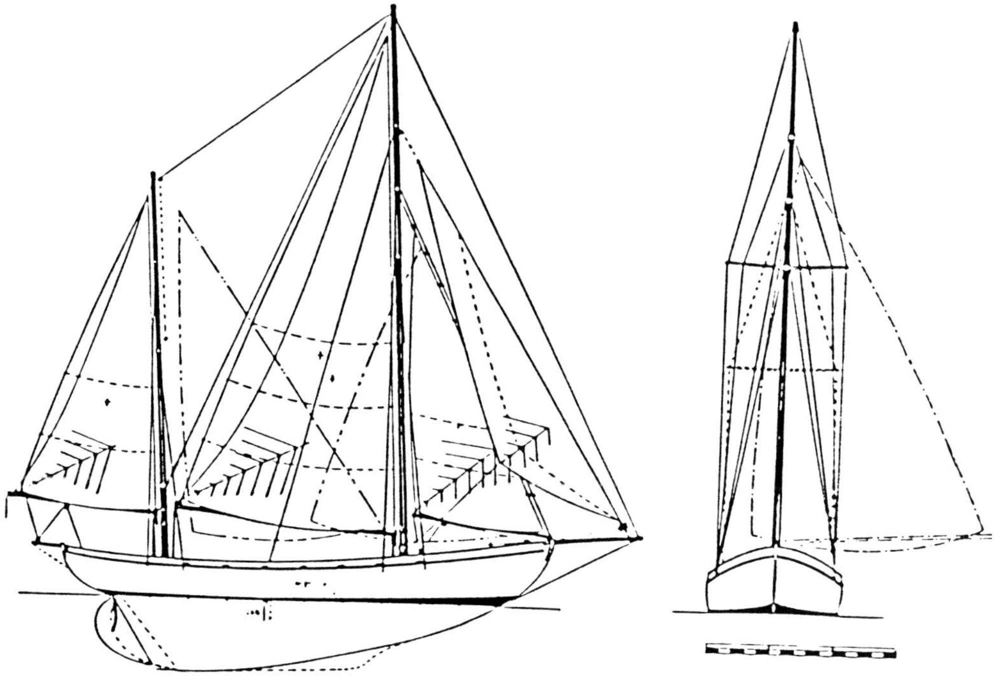

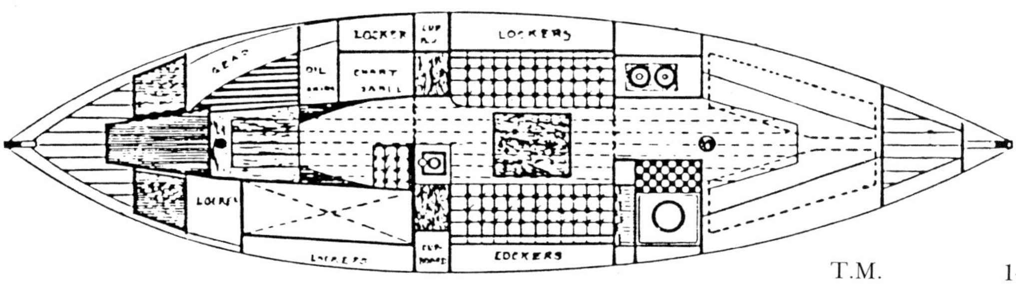

T.M.	14 tons
L.O.A.	38 ft.
L.W.L.	30 ft.
Beam	10 ft. 3 in.
Draught	6 ft. 2 in.
T.S.A.	662 sq. ft.

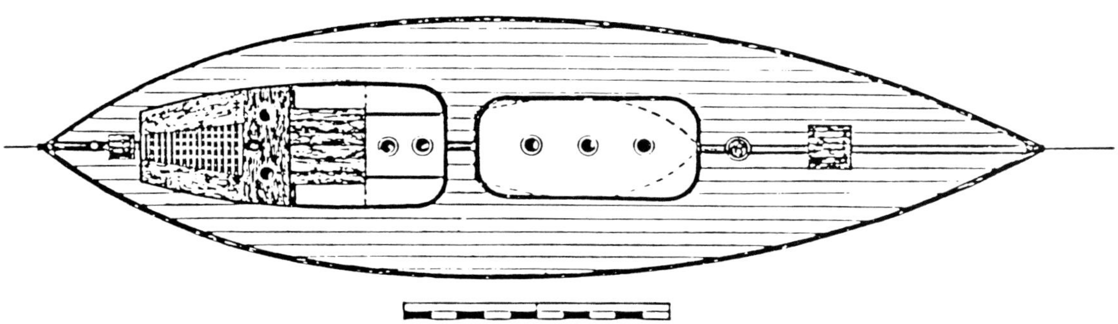

QUEEN OF ARDEN

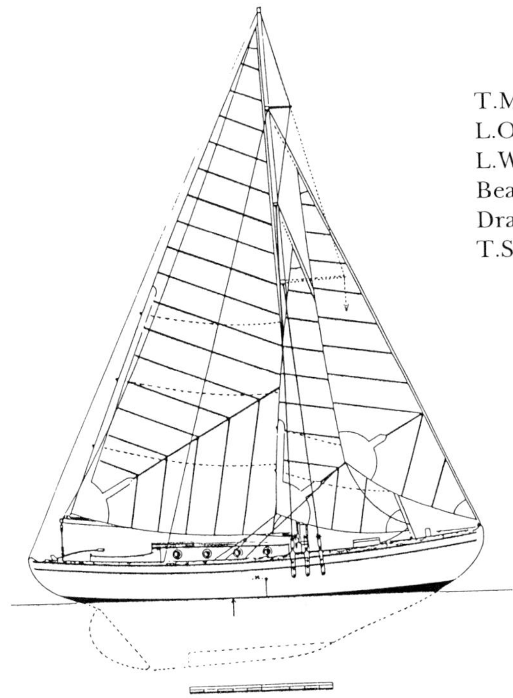

T.M.	8.3 tons
L.O.A.	32 ft.
L.W.L.	23 ft.
Beam	8 ft. 9 in.
Draught	5 ft.
T.S.A.	460 sq. ft.

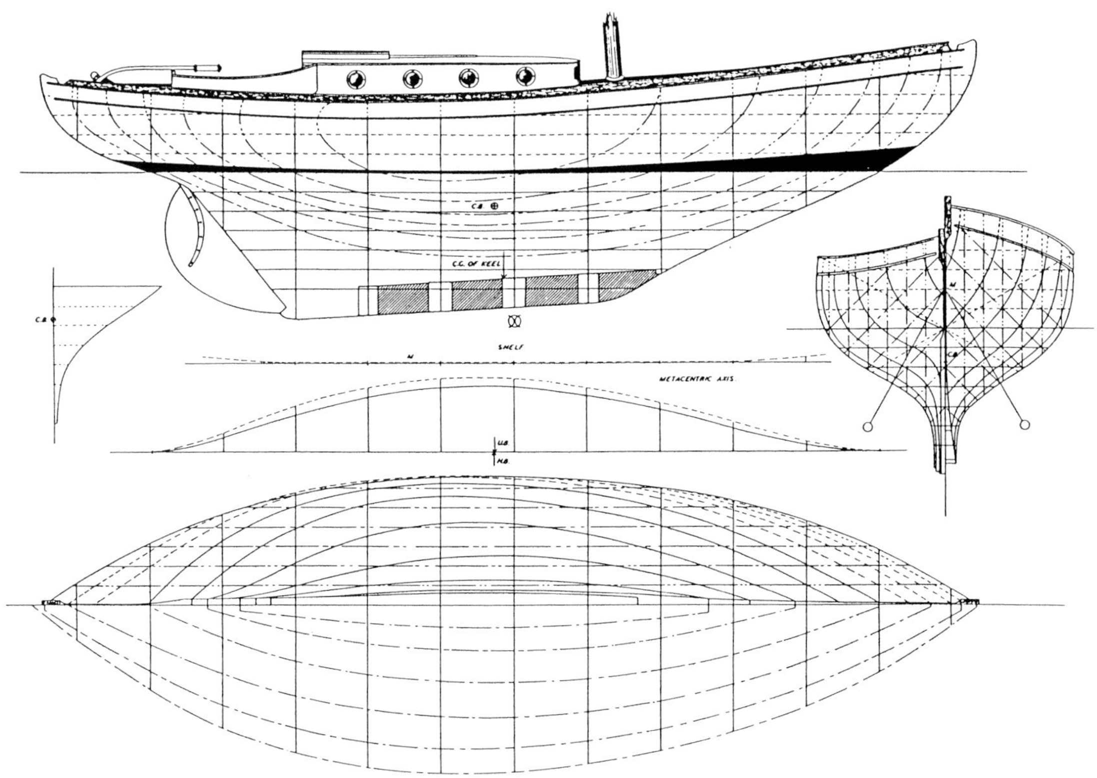

SYLPH OF ARDEN, THB's Penultimate Design

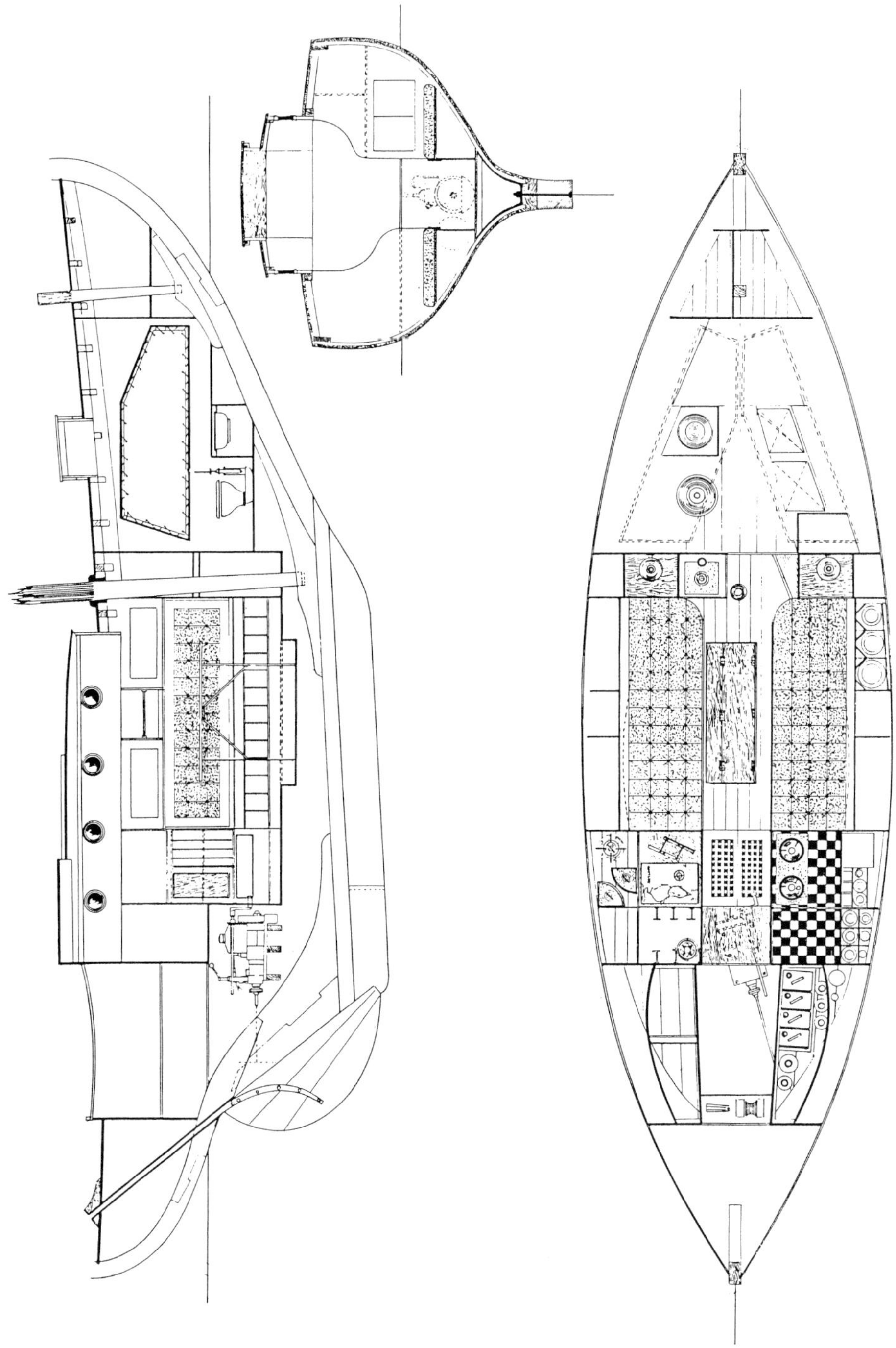

SLYPH OF ARDEN

THOMAS HARRISON BUTLER – A BIOGRAPHICAL PORTRAIT
by Joan Jardine-Brown

IT is not given to many people to be as widely known for their hobbies as in their professions. Thomas Harrison Butler was one such for, although most people nowadays know of him as a designer of small family cruising yachts, this was a leisure activity in which he was an amateur, albeit a very gifted one. He earned his living as a hard-working ophthalmic surgeon whose reputation as an innovative ophthalmologist stretched far beyond the shores of the United Kingdom, just as yachts have been built to his designs in many countries around the world. Even as I write, on New Year's Day, 1994, there are HB boats being built in Australia and here in Britain. He made no charge for his designs other than one guinea (£1.05p) for expenses incurred but he invited donations to whatever charitable cause he was supporting at the time.

One beneficiary was the church which he and my mother attended and on one occasion, when he was reading St Paul's account of the storm in chapter 27 of the Acts of the Apostles, during which four anchors were cast over the stern, and the crew undergirded the ship, he inserted his own description of the procedure, in case the congregation did not understand what was meant by 'undergirding'.

A biography should, however, start at the beginning. Butler family records go back a very long way but there is no mention of any seafarer until the birth in 1777 of Charles William Butler, the youngest son of the Reverend Weeden Butler (1742–1823) who was my father's great great grandfather. Charles was Captain of the East Indiaman, 'William Pitt', but had the misfortune of going down with his ship on January 17th 1814, when she was embayed in the notorious Algoa Bay in South Africa, during a westerly gale, and all hands were lost.

We are descended from the eldest son, the Reverend Richard Weeden Butler, who assisted and later, succeeded his father as Master of the fashionable private Classical School which they kept at No.6 Cheyne Walk, Chelsea, in London. Isambard Kingdom Brunel spent some time there as a pupil and, after the school was given up, our own branch of the Butler family continued to live in the house into the lifetime of my grandfather.

Richard Weeden's younger brother, the Very Reverend George Butler, D.D., who was Headmaster of Harrow for twenty-four years and afterwards Dean of Peterborough, headed a dynasty of intellectually gifted

Butlers which included a third cousin of T.H.B., the late 'Rab' Butler, former Cabinet Minister and Master of Trinity College, Cambridge, who became Lord Butler of Saffron Walden. Ours is the senior but less distinguished branch of the family.

At the time of my father's birth, his father was curate in St Thomas's Church in Stanhope, serving under the Reverend George Harrison, hence Thomas's second name. About two years later they moved to Dorset where his father had become rector of Broadmayne with West Knighton and Broadmayne Rectory was the home in which he grew up, close enough to Portland and Weymouth to engender an interest in and love of ships and the sea which lasted throughout his life.

He lived with his eccentric, Calvinistic father, his mother (whom he disliked), his younger brother (who became a clergyman) and his two younger sisters, in a stultifying, puritanical religious atmosphere which left its mark but happily, not its bigotry. He was scrupulously honest and truthful to the point of tactlessness on occasions but he was also broad-minded in many ways. He had an alert mind and was receptive to new ideas when once he had assessed their worth. He held strong views and expressed them forcefully, whether of approval or disapproval. He used to denounce vigorously the gaff-rigged fishing-boat types of craft which carry strong weather-helm, a fault which, after many years of striving, he managed to eliminate almost completely from the yachts built to his later designs, particularly those of his 'metacentric' period.

He was educated, first in the village/church school where he would have been taught by his father. He went next to Dorchester Grammar School from where, after having been failed in spelling and handwriting in an entrance examination for the Royal Navy, which he sat at a very early age, he gained an exhibition to St Paul's School in London.

At St Paul's he was well grounded in Greek and Latin as well as in the Sciences. He took an active part in the debates of the Philosophical Society, of which he became Secretary for a time. In 1889 he read a paper on 'marine engines' and on another occasion in the same year he discussed the probable uses to which aluminium might be put, should the cost of production be reduced, as seemed likely at the time.

Diaries written between 1889 and 1893 cover his last two terms at St Paul's, his three undergraduate years at Oxford and the start of his clinical years at St Bartholomew's Hospital in London. Probably, thereafter, he was too busy to continue the diaries. As his horizons broadened, at boarding school and university, meeting people from far afield and staying with friends, and relatives, one can sense a growing disenchantment with life in the suffocatingly narrow evangelical atmosphere of his home.

He rowed in his College Eight at Oxford and he continued to play cricket and tennis into adult life and was a frequent spectator at Warwickshire County Cricket Club's matches.

Sailing was first experienced when he was taken out by local boatmen from Weymouth, where visits were made for such diversions as dental treatment and the day enlivened by going to the Nothe to look at the schooners at anchor. Often, the rectory tricycle would be used for transport to Dorchester, to catch the train to Weymouth.

Mention is made in the diaries of playing the organ for services and comments made on his father's sermons. He also described enjoyable visits to Taunton to stay with his great-aunt Betsy, where he met his second cousin Ellen Reed, whom he subsequently married, after a six years engagement. Her father was a general medical practitioner in Westbury, Wiltshire. Fun was had in Taunton but then there was the inevitable return to the dreary rectory.

I knew my father as a very hard-working and industrious man and the diaries show that this had lasted from boyhood. Hours worked per day and per week and the averages were recorded, both in term-time and the holidays, as also were the winter temperatures in his room. Comfort was probably regarded as sinful.

His industry brought academic rewards: a scholarship from St Paul's to Corpus Christi College, Oxford in 1889, where he graduated with a First Class Honours degree in Natural Sciences, despite having led a full life, with a range of social activities and a good deal of research into yacht-designing at the Bodleian Library and Radcliffe Camera. In fact, what were probably his first attempts at designing were made during this time.

In 1895, at the end of his student clinical years at St Bartholomew's Hospital, he qualified as B.M., B.Ch.(Oxon) and M.R.C.S.(Eng.) and L.R.C.P.(Lond). He was awarded a Radcliffe Travelling Fellowship from Oxford in 1896 and this enabled him to further his post-graduate studies in Berlin, Kiel, Dresden, Vienna and Paris. He acquired a working knowledge of French and German which proved useful in his professional life.

Having completed his European studies, my father set sail aboard the S.S. 'Raglan Castle' for Port Elizabeth, in South Africa, leaving his fiancée behind, to join him as soon as he was earning enough to keep a wife (£300 per annum). They were married in 1900, in Cape Town Cathedral on the day her ship berthed.

Not a great deal of information about the time in South Africa is available but T.H.B. mentioned in one of his articles that he had built a skipjack to an American design and that she was a failure. He became Plague Medical Officer in Cape Town and drew upon his experiences of the widespread epidemic of bubonic plague for his doctoral thesis for which he gained the D.M.(Oxon) in 1902. The other noteworthy event was the birth of their first child, Rupert. My mother returned to England with Rupert when he was not much more than a year old, leaving my father to rejoin her later. There then followed a fairly short period in England, with my father in general practice in Westcliff-on-Sea, in Essex, but at this time he became interested in ophthalmology. He had a model of a head into the eye-sockets of which he placed pigs' eyes in order to practise operations. Rupert, then a very small boy, wanted to give his father pigs' eyes for a birthday present and was most indignant when an elderly great-aunt suggested that what he meant were (peppermint) 'bulls-eyes'.

Eric was born in 1902, before the family again went abroad, this time to live in the St John Ophthalmic Hospital in Jerusalem where my father was Assistant Surgeon. Nora was born in Jerusalem in 1904 and *The Yachtsman* printed a supplement in June of that year for a design which T.H.B. had sent, almost certainly the first of a very long line of his designs to be published. It had been made under great difficulties for the heat of the sirocco had warped his drawing-board, he had but half a spline and only a few very inaccurate French curves. He had been obliged to draw the curves freehand and ink them in bit by bit. This was done with his customary care and patience and the finished drawings show no sign of the struggle which had gone into their making. Long after they left Palestine, some Arabic words were remembered and were used for the names of designs and, of boats: *Sandook* (a box or, coffin) and *Moyezerka* (blue water or, a cataract of the eye).

Jerusalem marked the end of the family's peregrinations and the return to England was final. T.H.B. had worked his passage as Ship's Doctor on at least one of his voyages. It was not from choice that he spent the rest of his life living in Warwickshire, about as far from the sea as it is possible to be in England.

The post of Honorary Ophthalmic Surgeon (the equivalent of Consultant Surgeon but unpaid) in the Coventry and Warwickshire Hospital was offered and he was available to take it. This was a new appointment and he created the Eye Department in which he worked until he was not far off seventy-four years old. There followed similar appointments to the Birmingham and Midland Eye Hospital and to the Warneford Hospital in Leamington Spa. After retiring from the Warneford Hospital, and from the B.M.E.H. at the age of sixty, he was taken on at West Bromwich and during World War II, he returned to the B.M.E.H. to take the place of younger surgeons who were on active service.

The family lived first in Kenilworth and then moved to Leamington where my sister Cynthia was born, twelve years after Nora, and I was born twenty months later to complete the family. After Leamington we lived for about thirteen years in Birmingham but in 1935 we moved to Hampton-in-Arden, which explains the 'of Arden' suffix to some of T.H.B.'s designs of this period. My father travelled to work by train until the General Strike in 1926 when he bought a 'bull-nosed' Morris Cowley – and drove thereafter.

He died before the introduction of the National Health Service in 1949 and his hospital appointments provided no income but took up most of his working hours – and the ones he most enjoyed. However, with a family of five children to be fed and educated – no state help in those days – he required a private practice as well and he had consulting rooms in Birmingham, Coventry and Leamington. He drove many miles each week and worked long hours. He loved his work and never wanted to retire and said that he 'didn't want to sit decaying in a chair'. In fact, he remained in harness, working full time and operating until about three weeks before his death, and his eyes were as sure and his hands as steady as ever. His last designs show no trace of deterioration in his customary superb draughtsmanship.

Indeed, the drawing-boards were seldom unoccupied and nearly always there was one propped up on a bookcase in his study. I have my father's notebook in which the 'Details of Designs' are listed. Fifty-two designs are mentioned, starting with 'No.100 "Hong Kong" design perfected'. The vital statistics vary little from the design published in the June 1909 *Yachting Monthly*, of which T.H.B. wrote in the May 1911 issue that several boats had been built to the design, in Hong Kong and elsewhere. At least three were built in Hong Kong and formed a one-design class. These must have been the first HB boats to be built and I have a rather faded photograph of one sailing. The word 'class' is inapplicable to HB boats which are built to 'designs'. The nearest to a class are the Z.4-tonners which were the Zyklons built by Alfred Lockhart (Marine) Ltd. At least fifty-two of these were built. My father had no connexion with the project other than that his design was used. All Z.4s are Zyklons but not every Zyklon is a Z.4-tonner: many Zyklons were built singly, in various countries.

He made one or two models from his designs, using stiff cardboard for the skeleton and strips of gummed brown paper for the planking. A sewing needle was used for splices in the rigging and my mother made the sails. They did sail but quickly became rather sodden. His article: 'How to make a Cardboard Model' was published in the *Yachting Monthly* of April 1916.

Whilst recovering from influenza at the age of seventy, he wondered suddenly how much Greek he remembered from his schooldays, so collected his Greek New Testament from the study and went back to bed. He was very surprised by how much he did remember and thereafter, detective stories, which hitherto had been his main leisure reading matter, were supplanted by the works of Homer, Xenophon, Herodotus and Thucydides. He very much enjoyed his classical studies, sorting out the niceties of Greek grammar, which he perhaps did not appreciate fully when it was a compulsory subject at school. Suddenly, yacht designs began to sport Greek names.

Late in life, he was given two honorary awards, both of which gave him great pleasure. In 1938, he was made an Associate of the Institute of Naval Architects (now, the Royal I.N.A.) and, in 1941 he was made a Fellow of the Royal College of Surgeons of England.

Despite the many and varied interests which occupied him, he always had time for his family and, as children, we were treated to nightly instalments of an improbable story which went on for years. Grandchildren, too, were entertained by his imaginative stories and drawings. He was 'Tommie' or 'Thomas' (never 'Tom') to all of us: kind and generous, not only with financial help but also with his talents and his time. I do not remember him ever losing his temper though he could be rather sarcastic when he was annoyed. His tastes were simple: he was very unworldly and did not expect guile from others. He had no interest in money as such and never had a great deal, particularly as he often reduced his fees or, waived them entirely, but he was always ready to make an advance on one's allowance if asked. 'Everything I have is yours, dear', was the usual reply.

He was both critical and encouraging, setting high standards for himself and expecting them from others. He was meticulous in his work and in his drawings and had infinite patience and he was thorough in whatever he undertook. He disliked sloppy speech and bad grammar and was fascinated by words and their derivations. Life at home was interesting and stimulating with plenty of visitors who came to talk about boats – or eyes. Operations were frequently discussed at mealtimes and anatomical demonstrations given if for instance hearts were on the menu. We grew up not to be squeamish. One of his maxims was, 'Eschew anything which people say is good for you: it's bound to be nasty.' He was placid and good tempered; my mother was the fiery one.

He loved sailing, not only for itself but as an intellectual exercise, observing the behaviour and performance of the boat in different weather conditions and under different sail combinations. He sailed in 'his' boats whenever possible and was always critical of their faults which often had not been noticed by their owners. Weather-helm was something he deplored and when he found it, as in the Cyclone IIs, he went home and set to work on another design which he hoped would be an improvement.

There followed a sequence of 22′6″ LWL designs with transom sterns but, whereas Cyclone II was an enlargement of his 1919 19′ LWL Cyclone design, Yonne was practically an enlargement of his winning design 'No.16' (the competition entry number) in the 1926 *Yachting Monthly* competition. 'No.16' had an 18′ LWL. T.H.B. called Yonne the 'sports model' and described her as a single-masted staysail schooner!

Englyn followed on from Yonne and later, her sections were spaced out to give a 25′ LWL design, Askadil. Omega was the final design in the 22′6″ LWL series, the metacentric version, and Irmiger was the metacentric version of Askadil. Throughout the sequence, bow sections were filled out, quarters fined down and transoms became smaller – and balance was improved. Omicron was a reduced version of Omega, a 3-tonner with a 17′ LWL. The shape of the forefoot was altered to keep the weight low down.

Having decided that his own *Vindilis* had shortcomings, he embarked upon improvements and Edith Rose left the drawing board. He had increased the waterline by 6″ and replaced the counter with a canoe stern. Thereafter, most of his larger designs were given canoe sterns. He found that *Edith Rose* was a little tender but he had no complaints about her balance. However, the urge to improve upon the design resulted in Rose of Arden in 1938 and, about six years and many designs of varying types and sizes later, his last design, Dream of Arden.

Sinah, designed in 1935, was the first of his designs to be analysed when, in 1936, Eng. Rear Admiral Alfred Turner used it to show my father his method of determining the metacentric balance of a hull. She was found to be correct, metacentrically speaking. T.H.B. had achieved his goal of perfect balance without outside aids but thereafter, he had a means of verification and his subsequent designs were all subjected to metacentric scrutiny and adjusted if necessary to produce balanced hulls and indeed, well-balanced yachts resulted, as I have learned from their owners. Nowadays, computers have probably taken over but, for those without such aids, metacentric analysis is a very helpful guide to enable one to achieve equal buoyancy in the dissimilar ends of a design, at bow and stern.

Before he owned his own boat, T.H.B. used to cruise with his friends and colleagues, Claud Worth and Devereux Marshall, aboard the latter's ketch, *Maud* but, in 1912 he bought *White Heather* (ex-*Sapphire*), a half-decked Plymouth Hooker which he had decked in and made habitable and re-named, *Sandook*. We were all trained on board and she had no engine until 1933. She was sold at the end of the 1934 season and the sorrow of parting with such an old friend was tempered by the excitement of having *Vindilis* built. At long last, in 1935, my father owned a boat built to one of his designs. After a few teething troubles had been corrected she gave us some very enjoyable cruising until World War II intervened and, because all the family had married and left home, she was sold in 1939. His intention was to have a Z.4-tonner after the war but his death in 1945 made it but a pipe-dream. Sadly, the first edition of *Cruising Yachts* was not published until after he had died and I read the proofs and made the index.

Most of the first hundred designs no longer exist but a few remain in the early numbers of the yachting journals and elsewhere. He did not keep designs when they became obsolete. Albert Strange, who died in 1917, had an influence on my father's early designs and they used to correspond, and to meet occasionally when he came to have his eyes examined. HB designs are readily recognisable by his distinctive style of draughtsmanship and by the HB personality which permeates them all. They are works of art in their own right and are full of interesting details which indicate where articles might be stowed, such as, the lead-line, navigation lights (oil) and in one design, a chicken in a food-locker, but it is the flowing curves and the pronounced sheer which are the dominant features. The boats themselves are always a talking-point in an anchorage, where they stand out from the modern sharp-nosed angular GRP vessels.

His aim was to design family boats which would be easy on the helm, sea-kindly and seaworthy, have a reasonable turn of speed and be efficient, comfortable and good-looking. Being a cruising man himself, he knew what was required. He took care that seats were not too high and that there was adequate sitting headroom. In designs with a LWL of 22′6″ or more, there is 6′ headroom under the cabin-top. He would never destroy the appearance of a small boat by adding a dog-house. He was 6′ tall himself – and large, but had no qualms about sailing in a Zyklon. He had a particular flair for getting a quart into a pint pot down below. He abhorred wasted space and designed plenty of lockers – a place for everything, giving an impression of uncluttered space, even in the smallest boats. He was happiest when designing boats of sizes in which he had sailing experience and, because Maid of Arden fell outside this parameter, he invoked the professional aid of W. McC. Meek & Co. for the design of her lead keel and her construction plan.

He would be surprised to know that several HB boats have been race-winners, that two have sailed round the world and that many are making trans-oceanic voyages as a matter of course, some of them single-handed. He did not envisage such adventurous cruising for them.

The reasons for making designs were many but always, there was the wish for perfect balance in the resulting boat. Several designs were made for inclusion in the books of the late Francis B. Cooke, others were entries in designing competitions. When acting as a judge, he sometimes made a design himself, out of interest, embodying his own interpretation of the rules. Sometimes, he was asked to make a specific design but he would never include features of which he disapproved – an advantage of being an amateur designer. Mostly, he designed because he kept thinking of improvements to existing designs or to experiment with an idea, as in the case of Trutina in which the corresponding bow and stern sections are identical. He also made designs to illustrate articles that he wrote. Latterly, he designed fairly quickly and during the war years between 1939–1945, when he was particularly busy, and running his son Rupert's practice as well as his own, he made at least ten designs. He never stopped and, shortly after he died, the sail plan of Sylph of Arden came back from being duplicated.

I have striven, in this condensed biography, to portray the personality of one of whom I have the fondest memories. He will be remembered by his boats for as long as they last and, in 1973, the Harrison Butler Association was formed: a tribute to my father which I hope will continue for many years to come.

O. Joan Jardine-Brown
1994

INDEX